# The Foundations of
# Helicopter Flight

T0249510

**Frontispiece.** Westland Navy Lynx HAS 3 hovering above the flight deck of the helicopter support ship *Engadine*. Note the circular grid in the deck which is part of the helicopter's harpoon decklock system used to secure it to the ship's deck after touchdown. (Courtesy Westland Helicopters.)

# The Foundations of Helicopter Flight

*Simon Newman* BSc (Hons), MSc (Eng), CEng, MRAeS, FIMA, CMath
Lecturer in Helicopter Engineering, Department of Aeronautics and
Astronautics, University of Southampton, Southampton, UK

**B**UTTERWORTH
**H**EINEMANN

OXFORD  AMSTERDAM  BOSTON  LONDON  NEW YORK  PARIS
SAN DIEGO  SAN FRANCISCO  SINGAPORE  SYDNEY  TOKYO

Butterworth-Heinemann
An imprint of Elsevier Science
Linacre House, Jordan Hill, Oxford OX2 8DP
200 Wheeler Road, Burlington, MA 01803

First published 1994
Transferred to digital printing 2003

Copyright © 1994, Simon Newman. All rights reserved

No part of this publication may be reproduced in any material form (including
photocopying or storing in any medium by electronic means and whether
or not transiently or incidentally to some other use of this publication) without
the written permission of the copyright holder except in accordance with the
provisions of the Copyright, Designs and Patents Act 1988 or under the terms of
a licence issued by the Copyright Licensing Agency Ltd, 90 Tottenham Court Road,
London, England W1T 4LP. Applications for the copyright holder's written
permission to reproduce any part of this publication should be addressed
to the publisher

Whilst the advice and information in this book are believed to be true and accurate
at the time of going to press, neither the author nor the publisher can accept any
legal responsibility or liability for any errors or omissionss that may be made

**British Library Cataloguing in Publication Data**
A catalogue record for this book is available from the British Library

**Library of Congress Cataloguing in Publication Data**
A catalogue record for this book is available from the Library of Congress

ISBN 0 340 58702 4

For information on all Butterworth-Heinemann publications
visit our website at www.bh.com

# Preface

Perhaps I was destined to pursue a career in rotorcraft, since I was born within half a mile of a famous post mill in Buckinghamshire. A romantic thought but far from the true reason which was that after graduating in mathematics I received only one sensible offer of employment. So on 2 September 1970 I arrived in Somerset, dressed in a new suit, and feeling rather overawed at the prospect of getting to grips with helicopters. Not only that, what were the engineers working on helicopters like? My puzzlement was not helped by my first summons to the Research Director's office. Whilst waiting for the incumbent to see me, I spotted a framed photograph of an early helicopter inside which was a youngish looking pilot apparently hanging on very tightly. There was a handwritten inscription underneath reading: ". . . in the beginning there was blissful ignorance". The pilot was one of the best known names in rotary wing flight, namely Raoul Hafner. Over the years this message has become clearer and it sums up the feelings and attitudes of early pioneers.

The engineers, pilots, and other colleagues I got to know during my career have been enthusiasts about aviation with a love for the aircraft both present and past. There is a pioneering spirit with a willingness to stick the neck out, always tempered with a deep understanding and respect for the potential dangers of bad practice and inferior workmanship. Helicopter design and production is a serious business demanding total trust and confidence among all those involved. It is to their eternal credit that this is achieved without stifling the *esprit de corps* which is so necessary when monotonous checking and testing procedures have to be completed with complete safety and accuracy.

The mind is a good filter, and in writing this I am aware that mainly the good times are remembered, with the other side discarded. However, although in 1970 I did not intend a career in helicopters, I can have no regrets. I was lucky enough to enjoy the camaraderie of colleagues of superior ability, technical excellence, but willing to share their thoughts and ideas, and without them I could not contemplate writing this book. I will always be grateful to them.

This book is intended as an introductory text and to approach the subject from a practical point of view. Any topic which requires a high degree of mathematical analysis will appear dry unless the application of the knowledge is not emphasised. This I have attempted to do, introducing each topic so that it builds on those already covered. Practical examples are included to enable the reader to see the subjects "in action" and I hope this will cement the techniques into the reader's mind and most importantly show that they are of real use. Certain areas are referenced since they are covered very well by other papers or books. There are several very good texts on helicopter design and I am intending that this book will supplement them by providing a different viewpoint.

As with all major pieces of work I must express my thanks to many people, my old colleagues and sparring partners at Westland Helicopters in particular David Balmford, my contacts at research establishments, Alan Jones and Ron Walker, and the armed services especially Lt Cdr Jon Rich and Chief Petty Officer Steve Newberry of RNAS Culdrose. I should also like to express my grateful thanks to several sources for permission to include valuable photographs, namely Julie Pilbeam and Alan Jeffries of Westland Helicopters Ltd, Mary Anne Semmelrock of Kaman Aerospace Corporation, Sikorsky Aircraft, Terry Arnold of Bell Helicopter Textron Incorporated and the Ministry of Defence. To my mentor at University, Ian Cheeseman, I express grateful thanks and happy memories, I hope he has the same. I was always a bit slow at times and luckily he had the patience and armoury to penetrate my skull. To Stephen Hewitt I owe a special debt of gratitude for introducing me to the power of desk-top publishing and to many hours of reading the various drafts of this book and trying to spot any English which may have been present! Finally to my family, my dear wife Stella for feeding and clothing me when I was beavering away at the computer, proof-reading and providing the essential support so necessary when you are out on your uppers, my mother Olga for still thinking that I am the greatest thing since sliced bread, and my late father William who saw me through most of this project but I hope can still see me now completing the job, on time. To them I dedicate this book.

Simon Newman
Winchester, 1993

# Contents

# Abbreviations

| | |
|---|---|
| ABC™ | Advanced Blade Concept™ |
| ACSR | active control of structural response |
| ASW | anti-submarine warfare |
| AUW | all-up weight |
| | |
| CG | centre of gravity |
| CF | centrifugal force |
| | |
| DAVI | dynamic anti-resonant vibration isolator |
| $D_{100}$ | aerodynamic drag at 100 units of speed |
| | |
| FFT | fast Fourier transform |
| | |
| $g$ | gravitational acceleration |
| | |
| IGE | in ground effect |
| ISA | international standard atmosphere |
| | |
| $L_{100}$ | aerodynamic lift at 100 units of speed |
| | |
| MPOG | minimum pitch on the ground |
| $M_{100}$ | aerodynamic pitching moment at 100 units of speed |
| | |
| NFA | no feathering axis |
| NOTAR™ | no tail rotor |
| | |
| OGE | out of ground effect |
| | |
| SA | Sud Aviation |
| SFC | specific fuel consumption |
| SHM | simple harmonic motion |
| | |
| TRLF | transmission loss factor |
| | |
| vs | versus |
| VTOL | vertical take-off and landing |
| | |
| WG | Westland Group |

Aerospatiale Alouette III of the Grasshoppers display team.

# 1

# Introduction

*And I looked, and, behold, a whirlwind came out of the north, a great*
*cloud, and a fire infolding itself, and a brightness was about it . . .*

Ezekiel, Chapter 1

For many years news bulletins have shown helicopters operating in many
different situations all over the world. From being truly aggressive as seen in the
Vietnam war to the life-saving missions flown over stricken shipping; from
pleasure flying to local air ambulances; from supporting climbing expeditions in
the mountains to lifting and placing the spire on top of Coventry cathedral. The
helicopter is indeed a versatile aircraft.

It differs in appearance markedly from a fixed wing aircraft but must operate
according to the same aerodynamic laws. Make no mistake, whilst the helicopter
may, at times, appear rudimentary it is in fact a sophisticated vehicle which
imposes great demands on its designers.

This book is an introductory examination of these aircraft and the journey
encompasses their design, construction and operation. The methods are illustrated
with examples to show how and, very importantly, why they are of use. The
intention of the book is to remove the veil from this family of aircraft known as
helicopters and to show them in their true light. Whichever way you look at them
they are a unique class of flying machine.

The ability to hover for a period of time characterises, by definition, vertical
take-off and landing (VTOL) aircraft, and in this class of vehicle, the helicopter is
the main protagonist. At first sight the ability to hover may not appear to hold too
much of a difficulty. A jet engine should suffice to lift the airframe off the ground
and keep it there. This demonstrates the principle of hovering flight but does not
provide an aircraft capable of sustained hovering without the crippling penalty of
a high fuel consumption. Efficient hover comes from giving a large volume of air
a small increase in velocity and it is for this reason that the helicopter with its large
lifting rotors is the prime candidate for such use. However, in addition to hovering, the helicopter must be able to transfer into forward flight and back into hover,
efficiently and always under strict control. This puts a tremendous burden on the
helicopter since the requirements of hover and forward flight often provide con-
flicting demands on the airframe and rotors. The delicate balance of compromise
plays an important part in the formative years of the design.

Although it has similarities to a propeller, as it is a thrust producing device, the

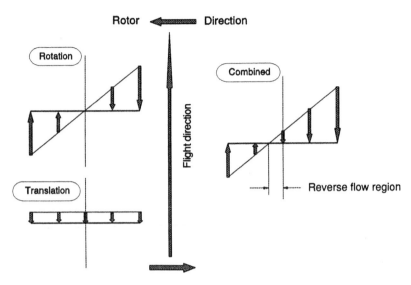

**Fig. 1.1.** Effect of forward flight velocity on helicopter rotor.

helicopter rotor has an essential difference. The propeller always operates in an axially symmetric environment, since it translates in a direction along its axis, and each blade will basically experience a constant situation. The interference of fuselage, wings, and engine nacelles will of course provide perturbations. This is not the case for the helicopter rotor. In hover, the axial symmetry observed by the propeller is also enjoyed by the rotor blade (with perturbations caused by the presence of the fuselage). However, as the helicopter commences its transition into forward flight, the rotor begins to move in an edgewise direction and the axial symmetry is lost. As the forward flight speed increases, this movement away from axial symmetry becomes greater.

Figure 1.1 shows the effects of this dissymmetry when the rotational inflow velocity seen by a rotor blade has a forward velocity added to it. This figure also shows the usual convention for expressing rotor blade azimuth position ($\psi$). The zero datum is when the blade points directly over the tail. (The forward flight direction is $\psi = 180°$.) The azimuth angle $\psi$ is measured from the datum in the direction of rotor rotation.

The rotor disc divides into two halves: the *advancing side* $(0° \leqslant \psi \leqslant 180°)$ is where the rotor blades move in the same direction as the airframe and the *retreating side* $(180° \leqslant \psi \leqslant 360°)$ is where the rotor blades move in opposition to the airframe.

As shown in Fig. 1.1, the advancing side will see the sum of the two inflow velocities and therefore an increased lift potential via the increased dynamic pressure. Conversely, the retreating side will see the difference of the inflow velocities, and its lift potential is therefore decreased. Indeed, Fig. 1.1 shows that a part of the retreating blade has the incident flow moving in a direction from trailing to leading edge. This is the *reverse flow region* which is circular and has its centre on the retreating blade at $\psi = 270°$ and passes through the rotor centre.

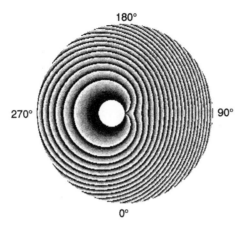

**Fig. 1.2.** Lift contours on untrimmed rotor in forward flight.

Figure 1.2 shows the contours of lift potential across the rotor disc in forward flight. It shows the dissymmetry of dynamic pressure due to the manner in which the rotation inflow combines with the forward flight speed. Clearly, if the blade aerodynamic angles were kept uniform around the disc, as in a propeller, the lift would be greater on the advancing side than the retreating side and a rolling moment would be set up. Additionally, this out of trim condition will cause high vibrations due to the oscillating loads in the blades as they rotate around the azimuth.

Juan de la Cierva in his early experiences with autogyros encountered this rolling moment effect and devised a solution which still exists today. His previous work as a structural engineer introduced him to pinned frameworks, which allow loads to be supported but afford relief to the structure from bending stresses. Transferring this concept to the autogyro rotor blade, he incorporated hinges into the hub whereby the blades were allowed to move out of the plane of rotation in what is termed flapping motion. The rolling effect is then uncoupled from the fuselage, because of the hinges, but a further problem arises. Since the blades are now freely hinged, they will move under the combined influences of aerodynamic and centrifugal forces, and as the rotor moves into forward flight, the rotor plane will in fact tilt backwards. The direction of the rotor thrust can be assumed to be perpendicular to the plane swept by the rotor blade tips so that any changes in the rotor disc plane are accompanied by a corresponding alteration in the rotor thrust line. For an autogyro, where the rotor is kept turning by an upflow of air through it, this rearward rotor plane tilt is what is required and the direction of rotor thrust force follows accordingly. A rearward component of the rotor thrust is therefore generated which, in the case of the autogyro, is overcome by the propulsive force given by the driving propeller.

With a helicopter, however, the main rotor thrust must provide, in addition to the vertical force supporting the aircraft, a forward force to overcome the aircraft's drag. The rearward tilt of the main rotor plane, used in the autogyro, is now in the wrong direction for a helicopter. Forward flight will require a forward disc tilt and

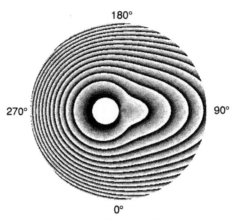

**Fig. 1.3.** Lift contours on trimmed rotor in forward flight.

to achieve it, whilst retaining flapping hinges, the pitch of the blades are altered cyclically as the rotor rotates with the intention of reversing the natural tendency of the rotor disc to tilt rearwards. Propellers have blade pitch change mechanisms for thrust control but all blades change pitch together (collectively). Helicopter rotors have also an adjustable collective pitch enabling the thrust to be varied, but to control the attitude of the main supporting rotor(s), as previously explained, will also require a cyclic pitch. The changing blade angle will influence the aerodynamics and the effect of the cyclic pitch on the lift potential contours of Fig. 1.2 is as shown in Fig. 1.3. As a comparison between Figs 1.2 and 1.3, contours of equal value intersect the centre line of the rotor ($\psi = 0°$ and $\psi = 180°$) at the same point. To trim the rotor in roll, the increasing lift potential of the advancing side has to be reduced with the opposite effect for the retreating side. This compromise, borne of the need to eliminate the rolling moment leads to aerodynamic limitations on the rotor, which have been a constant problem in the development of the helicopter rotor, particularly in the search for high speeds. The circular reverse flow region is shown in both Figs 1.2 and 1.3.

All lifting devices, be they wings or rotor blades, gain lift by superimposing a circulatory flow onto the flow passing the wing/blade. This causes a higher flow speed on the upper surface than the lower surface causing a pressure difference between them and causing lift to be generated. At the wing or blade tips this circulation must be shed into the flow as the load drops rapidly to zero. It does so by means of the well-known tip vortex often seen with high performance aircraft pulling tight turns in humid atmospheric conditions. The rapid rotation of the vortex causes a rapid chilling of the air in the central core of the vortex causing condensation to form and act as a marker, which is observed and shown in Plate 1.1. This also occurs with helicopter rotors as shown in Plate 1.2, but as a fixed wing aircraft leaves the vortices behind, the helicopter rotor blades, because of their motion, are forced to operate under the influence of the tip vortices shed by all the rotor blades (including themselves). Plate 1.2 also shows one blade passing close

**Plate 1.1.** Wing tip vortices generated by a McDonnell Douglas Phantom F4 in a tight turn.

**Plate 1.2.** Blade tip vortices generated by a Westland Sea King Mk5 in slow rearward flight. (© British Crown Copyright 1993/MOD reproduced with the permission of the Controller of Her Britannic Majesty's Stationery Office.)

**Fig. 1.4.**

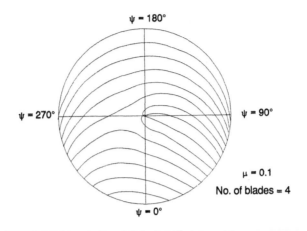

**Fig. 1.5.**

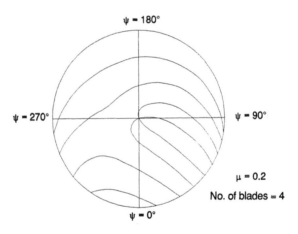

**Fig. 1.6.**

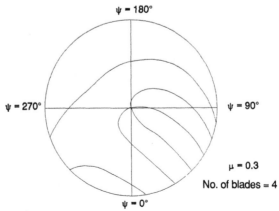

**Figs 1.4–1.6.** Vortex trajectories.

to the vortex shed by the immediately preceding blade. As each blade rotates it passes over or under vortices shed by itself and the other blades. Figures 1.4–1.6 show how these blade/vortex intersection points are distributed across the rotor for three forward flight speeds expressed in values of advance ratio ($\mu$)—see Chapter 5. The basic pattern can be seen and as each blade encounters a vortex, an abrupt change of aerodynamic loading is experienced. In this way the complicated environment of the helicopter rotor will cause major effects which influence the performance of the aircraft.

We are therefore faced with an aircraft capable of hovering and transferring to and from forward flight, fitted with hinges allowing the rotor blades to move out of the rotation plane, while being immersed in a hostile aerodynamic situation. The helicopter is a unique source of these problems and this book will describe these complications, explain why they occur, how they can be modelled and what conclusions can be drawn.

A misunderstanding sometimes occurs in the effect of engine failure on the ability of the helicopter to fly. It is a fact that should the pilot take no remedial action after an engine power failure the aerodynamic drag on the rotor blades will cause the rotor to slow, the rotor thrust to decrease and the aircraft to descend. Ultimately this will result in the rotor coming to rest with the inevitable loss of the helicopter. However, the ability of the air to turn the rotor is amply demonstrated with the autogyro and the essential fact is that the forward speed of the aircraft gives an upward component of the air through the rotor, because of its rearward tilt. This upward flow causes the lift forces on the rotor blades to change from slightly rearwards to slightly forwards and one component of the drag force on the blades becomes a propulsive force driving the rotor.

The helicopter can achieve this situation by the pilot lowering the collective pitch, hence reducing the main rotor thrust and allowing the aircraft to descend. The descending helicopter rotor now has an upflow of air through it and the rotor can be driven without the need of an engine. This technique of using the helicopter altitude to permit descent, under control, with the rotors turning is termed autorotation and is discussed in detail.

The single main and tail rotor helicopter configuration is the most common and, for this reason, this book will concentrate on it. Plate 1.3 shows a Westland Westminster in flight in its crane version which allows the components to be easily viewed. The main rotor supports the airframe and the main gearbox lies underneath. The main rotor controls can be seen surrounding the drive shaft. The fuselage is then attached to the gearbox by securing the lift frames to the gearbox feet. The tail rotor drive shaft can be seen coming from the rear of the main gearbox, passing along the spine of the tail boom, and via an angle gearbox, to the fin and finally to the tail rotor gearbox. The Napier Eland gas turbines are positioned on top of the cabin with the drive shafts connecting them to the main gearbox. The engines are contained within firewalls. At the base of the fin can be seen a bellcrank which forms part of the tail rotor control mechanism. This has to connect the pilot's pedals with the tail rotor pitch change mechanism and therefore has to run the full length of the airframe which is often achieved by means of cables. Although this particular aircraft was designed 40 years ago, the basic layout is typical.

**Plate 1.3.** Westland Westminster in its crane version. (Courtesy Westland Helicopters.)

In describing techniques used to design a helicopter, the complexity of the theoretical models can vary enormously. However, the book will start with the simplest and show the limitations of the method and the scope for improvement. More detailed theories will be described when they are necessary to understand important details.

A brief description of the chapters follows, but it should be borne in mind that the helicopter designer must find a solution which furnishes an aircraft capable of satisfying a mission requirement supplied by a customer and selling it with the obvious financial constraints this imposes. Changes to requirements whilst the design is under way will usually cause major modifications which are seldom understood or appreciated by the client. A good illustration of this is contained in two papers listed in the bibliography about the Westland WG13 project. It would be wrong to reference them immediately since the understanding of many of the descriptions and conclusions requires the reader to be familiar with the basic design principles of a helicopter. It is therefore recommended that they be consulted after the reader has assimilated the contents of the book.

The layout of the chapters is as follows.

### Chapter 2. Rotorcraft configurations

The vast majority of helicopter types are of a single main supporting rotor, internally driven, with a single tail rotor providing anti-torque control. Although this book concentrates on this type of aircraft, over the years there have been many different variants and this chapter describes the most common covering their control, with an indication of their relative strengths and weaknesses.

*Chapter 3. The rotor in hover*

The aerodynamic analysis of the main rotor begins in this chapter, where the axisymmetric flight condition of hover is analysed. Two main approaches are used namely, momentum and blade element theories. These two techniques are then developed in successive chapters.

*Chapter 4. The rotor in axial flight—climb and descent*

This chapter discusses vertical flight which retains the axial symmetry, but extends the analysis to include the axial inflow generated by a vertical speed. Whilst the momentum theory can be used to model all vertical flight conditions, and solutions exist, some conditions in low speed descent cause this approach to be unsuitable because of the physics of the situation. This breakdown in the applicability of the momentum theory is discussed and the physical reasons described.

*Chapter 5. The rotor in forward flight*

This chapter takes the main rotor into forward flight when the axial symmetry is lost. The variation of power with forward flight speed is derived and the introduction of cyclic pitch to the blade element theory explained. The need for roll trim, as indicated earlier in this chapter, and its effect on the rotor thrust limitations are described.

*Chapter 6. Dynamics of the rotor*

The rotor blade behaviour is treated in this chapter. Initially the blade is treated as rigid, but free to move relative to the rotor hub via rotation about hinges. The requirements for rotor hub construction in the form of additional hinges for different movements are detailed.

*Chapter 7. Autorotative performance*

The aerodynamics of the descending helicopter, initially introduced in Chapter 4, are analysed in more detail, particularly regarding its behaviour when power to drive the rotor is lost. The techniques used to achieve a safe take off or landing are described.

*Chapter 8. Blade dynamics*

Chapter 6 introduced the dynamic behaviour of the rotor blade, but only considered the basic motion when a rigid blade model can be used with confidence. The analysis of blade motion is extended in this chapter by considering the effects of blade flexibility. This is particularly important when vibration, or aeroelastic instabilities are under investigation.

*Chapter 9. Helicopter trim*

The rotor hub construction and the effect of the fuselage are introduced in this chapter. The calculations necessary to trim the helicopter longitudinally are derived for an elementary model and the required blade control angles are calculated using results derived in Chapter 6.

*Chapter 10. Vibration and its transmission*

Vibration is a constant problem because of the complicated aerodynamic environment in which rotor blades have to operate. The transmission of periodic forcings from the blades to the fuselage is described and how certain frequencies are filtered out. Methods to reduce vibration problems and their transmission to the airframe are described.

*Chapter 11. Tail rotors*

The tail rotor, because of its operational requirements, has problems of its own which differ markedly from those affecting the main supporting rotor. This chapter discusses these differences and how designers have avoided or reduced them. The chapter concludes with a discussion of aeroelastic phenomena peculiar to the tail rotor.

*Chapter 12. Ground resonance*

As described in Chapter 6, flapping motion requires not only a flap hinge allowing blade movement normal to the rotor plane, but also a lag hinge to relieve the Coriolis loads in the plane of rotation generated by the flapping motion. The existence of lagging motion brings with it the potentially catastrophic instability known as ground resonance. This chapter describes its origins, mannerisms and how it can be controlled. The phenomenon of air resonance is described.

*Chapter 13. Rotor speed governing*

The helicopter rotor is usually designed to operate at a nominally constant rotor speed. Earlier aircraft used a hand twist-grip throttle to control the engine(s), but in recent times rotor speed governors have been fitted to powerplants taking over this responsibility from the pilot. This chapter describes their method of operation and develops a simple mathematical model to demonstrate their characteristics.

*Chapter 14. Methods of calculating helicopter power, fuel consumption and mission performance*

This chapter collates earlier theories of helicopter performance, and presents them in a framework typical of a project office definition. It shows how the theories can be used to calculate the power required and hence fuel consumption of a specified helicopter design in any flight condition. This can then be used to "fly" a design through a mission, the aim of which is to allow parametric changes to the aircraft to be "flown" and the effects determined. The chapter concludes with the method used on two distinct missions. They are of a military nature, but are intended to show the differences in the requirements when a mission is predominantly high speed or hovering flight. Several parametric changes are made, the effects presented and conclusions drawn.

*Chapter 15. Epilogue—the windmill*

This concludes the book and reflects on various features of the helicopter as compared with the windmill, which has been developed over centuries without the use of computers but the cold light of experience. Remarkable similarities are seen and these are described.

The preface started with a windmill. It seems appropriate to close with a brief examination of the achievements of millwrights and why their ingenuity is so praiseworthy.

# Rotorcraft configurations

## Introduction

One of the simplest examples of a helicopter is the children's toy[1] known as the Chinese top. It consists of a propeller fitted to an axle which is spun with the hands. A development of this consists of two rotors on the same axle and rotating in opposite directions. The motive force is provided by a rubber band and when wound up under tension and released, the top will lift into the air under the thrust of the two rotors. Due to Newton's third law, the rotors will spin in opposite directions giving no net torque to the combined system. As well as showing the basic lifting principle of the helicopter rotor, it also demonstrates the anti-torque requirement of the combined system. As will be described in this chapter, the various rotorcraft configurations, which have been either proposed or flown, control this anti-torque requirement in different ways.

The essential requirements for any flying machine are:

(a) The lift force must equal or exceed the aircraft's weight and be capable of being controlled.
(b) The propulsive thrust in a forward flight direction must equal, or exceed, the aircraft's drag and be capable of being controlled.
(c) There must be control forces and moments, which are capable of altering the aircraft's attitude in pitch, roll and yaw.

Rotorcraft produce their lift force by means of a rotor or rotors, which are rotating wings, and consequently these rotors must be able to supply the above three

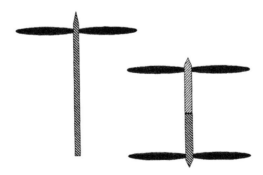

The Chinese top and its twin rotor derivative.

requirements. This means that forward flight can be sustained at speeds which are lower than those expected with a fixed wing vehicle. If the rotor is driven by means of extracting the driving energy from the air (the autogyro), then low speed flight is possible. However, if the rotor is driven by an internal engine (helicopter), then the low speed flight can indeed be extended to zero, enabling the aircraft to hover. Since the rotor of an autogyro extracts energy from an outside agency (i.e. the air passing upwards through it), there is no tendency for the aircraft to rotate in the opposite direction to the rotor. This still complies with Newton's third law because the torque balance is maintained by the air passing through the rotor inheriting a swirling velocity in the opposite sense to the rotor and thus maintaining the status quo. The development of the Chinese top, however, is an example of an internal power supply which will therefore require some anti-torque balancing (which is the other rotor). The helicopter has developed into many different configurations, and each one has its own method of torque balance.

It is necessary for the thrust on a main supporting rotor to be altered in direction during flight to give the required control. The direction of a rotor's thrust is usually taken to be perpendicular to the plane of the rotor itself (the plane traced out by the blade tips), making the thrust vector (i.e. magnitude and direction) controllable by varying the plane of rotation of the rotor in order to align its normal with the intended direction of the thrust force.

In the past there have been helicopter designs where more than two main rotors have been used, the Cierva Air Horse, for example, had three. However, in modern times, the usual number of main rotors, in a multi-rotor configuration, is two.

**Plate 2.1.** Westland Sea King Mk 5. Note the doghouse fairing behind the main rotor hub and the beanie cap above it. (© British Crown Copyright 1993/MOD reproduced with the permission of the Controller of Her Britannic Majesty's Stationery Office.)

## The single main rotor/tail rotor configuration

The single main rotor provides the vertical thrust necessary for the aircraft to leave the ground and control it in a vertical direction. Control in the fore, aft and lateral directions is also provided by the main rotor by altering the rotor disc plane and hence the direction of the thrust vector. This thrust vector change will also give the control in roll and pitch. With an internal power supply, yaw control is provided by a smaller rotor fitted at the tail end of the fuselage. It is positioned in a vertical plane and provides the horizontal force required to give the pilot control over the total yawing moment applied to the fuselage about the main rotor shaft axis. An example of this configuration is the Westland Sea King (see Plate 2.1).

The rotor control and trim of this aircraft are as follows:

| | |
|---|---|
| Vertical | main rotor thrust |
| Longitudinal | main rotor tilt fore/aft |
| Lateral | main rotor tilt lateral |
| Pitch | main rotor tilt fore/aft |
| Roll | main rotor tilt lateral |
| Yaw | tail rotor thrust/engine torque. |

## The twin main rotor configurations

Anti-torque control can be achieved by having two main rotors rotating in opposite directions, thereby cancelling out the torque from each other. The vertical control is shared between the rotors, and individually altering each rotor in thrust and direction will allow the aircraft to be controlled in the other five degrees of freedom (fore, aft and lateral movement; roll, pitch and yaw rotations). Twin main rotors, when placed at separate locations, allow the helicopter to be trimmed in flight with a much greater range of centre of gravity (CG) positions than the single main rotor variant will allow. The CG range allowable with a single main rotor configuration is dependent on the so-called control power of the rotor. This is the moment generated about the helicopter CG by a unit tilting of the main rotor disc plane. The moment depends on a combination of the rotor thrust and the hub construction. This is discussed in more detail in Chapter 9. Although there is some CG range with these helicopters, most of the twin main rotor configurations have a larger CG range (the coaxial is the exception), because trim can be obtained by unequal thrusts on the rotors.

### Tandem

The tandem configuration has a main rotor located at each end of the fuselage (see Plate 2.2). This means that the main rotors are aligned in a longitudinal direction. In order to allow the rear rotor to fly in as undisturbed air as possible, it is raised on a pylon above the level of the front rotor. This can give rise to problems when the aircraft is slowing down, when it will be in a nose-up attitude. The rear rotor will sink into the downwash of the front rotor and, unless more thrust is demanded of the rear rotor, the helicopter fuselage will tend to pitch to an even more nose-up

**Plate 2.2.** Westland Belvedere tandem rotor helicopter. Note the overlap of the rotor discs. (Courtesy Westland Helicopters.)

attitude. If a landing is being conducted, and the aircraft is at low altitude when the nose up flare is initiated, the aircraft is in danger of striking the ground with the tail section of the fuselage.

The rotor control and trim of this aircraft is as follows:

| | |
|---|---|
| Vertical | main rotor thrusts (collective) |
| Longitudinal | main rotors tilt fore/aft |
| Lateral | main rotors tilt lateral |
| Pitch | main rotor tilt fore/aft; main rotor thrusts (differential) |
| Roll | main rotor tilt lateral |
| Yaw | differential main rotor tilt. |

The longitudinal positioning of the rotors will allow a wide variation of CG position in a longitudinal sense. This will have a control drawback if the CG is not mid-way between the rotor centres. This means that the rotors will have different thrusts and if lateral control is demanded by a lateral tilt of the main rotors then a yawing moment will be created. In addition, the vertical separation of the planes of the rotors will give rise to problems in yaw. To turn the tandem helicopter in yaw, the rotor planes are tilted in opposite directions. This opposing sense of lateral rotor tilt will give rise to a rolling moment, which is in the wrong sense for a coordinated turn. For example, a port turn will be accompanied by a starboard rolling of the aircraft.

## Side by side

The side by side configuration has the main rotors placed in a lateral sense on pylons, i.e. perpendicular to the tandem (see Plate 2.3).

The rotor control and trim of this aircraft is as follows:

| | |
|---|---|
| Vertical | main rotor thrusts (collective) |
| Longitudinal | main rotors tilt fore/aft |
| Lateral | main rotors tilt lateral |
| Pitch | main rotors tilt fore/aft |
| Roll | main rotors tilt lateral; main rotor thrusts (differential) |
| Yaw | differential main rotor tilt fore/aft. |

The spacing of the rotors will allow a wide variation of CG position in a lateral sense. This is not nearly as useful as a longitudinal variation but if weapons are being deployed in a lateral sense, they can be released asymmetrically with no

**Plate 2.3.** Mil-Mi 12 side by side rotor helicopter which in August 1969 set a world payload record by lifting a massive 40 204.5 kg. (Courtesy Westland Helicopters.)

**Plate 2.4.** Kamov Ka-26 coaxial rotor helicopter. (Courtesy Westland Helicopters.)

danger of roll control being compromised. The lateral disposition of the main rotors means that the mutual interference is substantially reduced.

In forward flight, the rotors will experience the same aerodynamic conditions,
the absence of which is a problem with the tandem configuration through the rear
rotor flying in or close to the turbulence and wake of the front rotor. A disadvantage with the side by side configuration is the extra drag of the fuselage caused by
the supporting pylons.

## Coaxial

The coaxial configuration has the two main rotors on same axle. (This is like the
development of the Chinese top. See Plate 2.4.)

The rotor control and trim of this aircraft is as follows:

| | |
|---|---|
| Vertical | main rotor thrusts |
| Longitudinal | main rotors tilt fore/aft |
| Lateral | main rotors tilt lateral |
| Pitch | main rotors tilt fore/aft |
| Roll | main rotors tilt lateral |
| Yaw | differential main rotor torques. |

This configuration has the advantage of compactness which explains why the
coaxial helicopter is often seen on board ships. However, because of smaller

rotors giving a higher disc loading and the lower rotor flying in the downflow caused by the upper rotor the power consumption tends to be higher than other twin main rotor configurations. The linkages required to control both rotors on the same axle make for a bulky hub system with the attendant problems of aerodynamic drag. Yaw control is achieved by means of differential rotor torques, which can give rise to a problem in descending flight when the helicopter is in autorotation. The pedals, which are the pilot's yaw control inputs, are adjusted for correct behaviour in powered flight. That is a yawing moment is created in the direction appropriate to the pedal movement. In autorotation, where the lift forces on the blades change in direction to allow the air to drive the rotor, the yawing moment will now be opposite to the pedal movement resulting in a control reversal.

## Synchropter

The synchropter configuration has two axles very close together and inclined outwards. It is similar to the coaxial in its achievement of yaw control, but the rotors are now at the same height giving a more compact hub arrangement over the coaxial configuration. The problem of blade clearance means that the rotors are normally two bladed (see Plate 2.5).

**Plate 2.5.** Kaman H-43B Huskie synchropter. (Courtesy Kaman Aerospace Corporation.)

The rotor control and trim of this aircraft is as follows:

| | |
|---|---|
| Vertical | main rotor thrusts |
| Longitudinal | main rotors tilt fore/aft |
| Lateral | main rotors tilt lateral |
| Pitch | main rotors tilt fore/aft |
| Roll | main rotors tilt lateral. |
| Yaw | differential main rotor torques. |

The inclination of the main rotor axles will mean that there will be an element of cross coupling between the various rotation (*roll, pitch* and *yaw*) and translation (*longitudinal, lateral* and *vertical*) degrees of freedom. This inclination of the rotor shafts means that the rotor reaction to the engine torques will now have a component in the pitching axis. If the rotor rotations are such that the blades approach over the tail boom, the pitching moment obtained gives a stability to the aircraft with respect to vertical disturbances. In this case the rotors turn anticlockwise for the port rotor and clockwise for the starboard. Because of the outward tilt of the rotor shafts, the torque applied to the rotors will have a nose down component, hence the engine torque reaction is nose up which, in order to restore trim to the helicopter, requires the rotor disc planes to be tilted forward such that the rotor thrust vectors will now have a line of action which passes to the rear of the aircraft CG. If the helicopter encounters a vertical gust in an upward direction, there will be an increase in the rotor thrusts which means that the pitching moment on the helicopter will become more nose down and hence automatically correct the upward vertical translation of the aircraft caused by the thrust increase.

## *Compound*

All helicopters suffer from the criticism of a limitation in forward flight speed. There are various design changes which it is possible to use in order to increase the maximum speed of the aircraft. The thrust compounded helicopter has auxiliary propulsion installed (see Plate 2.6). As an example, the Fairey Gyrodyne used a single forward facing propeller for providing auxiliary propulsion (see Plate 2.7). It also was used as a replacement for a conventional tail rotor by providing the necessary torque reaction. The lift compounded helicopter has wings fitted to the fuselage with the aim of relieving the main rotor of the requirement to support the entire weight of the helicopter. This means that the main rotor thrust can be reduced with the intention of moving the rotor away from a stalling condition, which is the main aerodynamic limitation to the rotor's performance.

The problem posed by this solution is that the main rotor is still required to supply all of the propulsive force, and in order to overcome the drag of the helicopter the rotor will have to be tilted still further forward as the thrust magnitude has been reduced. To achieve this, more cyclic pitch has to be applied, which in the critical region ($\psi = 270°$) has the effect of almost negating the benefits of reducing the rotor thrust. A small increase in forward speed is possible (10 knots say), but nothing like the improvement one might expect at the outset. (The limitations to helicopter forward speed are discussed in Chapter 5.) The problem can

**Plate 2.6.** Sikorsky S-72, Rotor Systems Research Aircraft (RSRA). This is a compound helicopter used for rotor research which has sufficient wing area to support the aircraft enabling it to fly with the rotor operating at a much reduced thrust level. (Courtesy Sikorsky Aircraft.)

**Plate 2.7.** Fairey Gyrodyne. Past world speed record holder of 124 mph achieved in June 1948. (Courtesy Westland Helicopters.)

be avoided by divorcing the main rotor from the need to provide the propulsive as well as the supporting force. This enables the rotor to operate at a reduced thrust because of the lift contribution by the wings. This will mean the addition of extra propulsion and this configuration with both lift and thrust compounding is known as a fully compounded helicopter.

# Tip driven rotorcraft

The configurations described so far all have the rotors driven via the shaft. To do this, an elaborate transmission system must be used to reduce the high rotational speeds of the engine output shaft to that required by the rotors. The central part of this transmission is the main rotor gearbox which must be able to transmit the required torque and additionally provide the direct connection between the rotor(s) and fuselage. This requires the gearbox to withstand the flight loads placed on it to control the aircraft. For these reasons it is a substantial piece of engineering and can comprise up to 10% of the helicopter's overall weight. In the case of the single main rotor configuration the provision of shaft drive also requires a means of yaw control which is usually provided by the tail rotor, thereby further adding to the weight of the transmission.

If the main rotor was powered by a jet efflux emanating from each blade tip, then the main rotor gearbox and tail rotor system are not required and a substantial increase in payload can be achieved. A tail rotor can be fitted, but it is relieved of the need to overcome the torque of a shaft driven main rotor. A survey of helicopter development shows an almost total dominance of shaft driven rotors, so with the above possible savings in aircraft weight why has the tip driven rotor not been used more? The reasons seem compelling. Basically, the correct choice of rotor propulsion is dependent on the desired operation of the proposed helicopter design. In order to obtain sufficient torque to rotate the main rotor, a high jet efflux velocity is necessary since the blade tips are rotating at a considerable velocity and the jet must overcome this. If a low jet velocity is required then a correspondingly high mass flow through the duct must be obtained making for a large blade aerofoil cross-section.

The types of tip drive which can be used are explained below.

(*a*) *Pressure jet.* Gas is produced in the fuselage and fed to the rotor hub via rotating seals and then to the blades themselves. It is then passed through ducts within the rotor blade aerofoil section to the tips where it is turned through 90° and exits from the tip. This thrust can be augmented by a form of reheat where fuel is burnt at the tip. The pressure jet will require ducting in the blades which may cause conflict in the required mass flow, and therefore duct size, and the blade aerofoil section. Losses will occur in the duct but centrifugal pumping will help to offset these losses. If reheat is used then thermal shielding of the inner parts of the rotor blade may be necessary.

(*b*) *Tip mounted gas turbine.* A gas turbine jet engine is installed at each rotor blade tip, and fed with fuel via pipes in the rotor blades. However, the engine will have to work in a high centrifugal force field. For example, an engine installed in a fighter aircraft may be expected to withstand 10*g* accelerations whilst, in comparison, a rotor tip jet would be subjected to accelerations of the order of 500*g*. Additionally the spools in the engines will suffer considerable gyroscopic moments as the blades rotate and a twin spool arrangement with opposing directions of rotation will therefore be necessary.

(*c*) *Tip mounted pulse jet*. This can operate with a stationary rotor and has a lower fuel consumption than the ram jet, but the pulse jet has a highly stressed intake valve with only a limited life, and for optimum efficiency will require a lower rotor tip speed (exacerbating the retreating blade problem) and a long jet pipe.

(*d*) *Tip mounted ram jet*. This offers the simplest mechanical solution but, unlike the pulse jet, will not operate at zero rotor speed. This means that extra provision to spin the rotor up from rest will be necessary. It also suffers from a high fuel consumption. (For a subsonic ram jet a high tip speed is required of the rotor which will exacerbate the advancing blade problem.)

All of the above schemes suffer from the same disadvantages, namely:

(i)   Increased weight at the blade tips—requiring an increase in blade and hub strength.
(ii)  Increased drag at the blade tips giving an increase in profile power with the by-product that a higher descent rate will be necessary for autorotation (see Chapter 7).
(iii) Complications to the rotor hub and blades with seals and ducting.
(iv)  Noise generation—with the advent of stealth for military aircraft and noise pollution controls for civil operations, this is a major drawback.

Various analyses can be performed on the merits of shaft and tip driven rotorcraft. The major conclusions are that the tip driven rotors are best suited to large aircraft designs which, if shaft driven, would have a proportionally greater transmission weight. Also, they should only be operating in hover for limited periods where the fuel consumption is inferior to a shaft driven equivalent. A good example of this type of aircraft is the Fairey Rotodyne as shown in Plate 2.8. This had two Napier Eland gas turbine engines which provided a gas supply to the tip jets which were reheated. Tip drive was used in the hover but in forward flight the aircraft operated as an autogyro with the Eland engines powering two airscrews to provide the propulsion. The tip jet drive was throttled back as transition from hover to forward cruising flight was accomplished. The tip drive was progressively restored when the aircraft came back to the hover. Having attended a school not two miles from the Farnborough Air Show in the late 1950s I can personally testify to the noise problem!

# Advanced types of rotorcraft

## *Advancing Blade Concept (ABC™)*

This is essentially a coaxial configuration,[2] however the blades and hub are very stiff and the rotors can fly deliberately out of trim without clashing blades.

**Plate 2.8.** Fairey Rotodyne tip driven compound helicopter. (Courtesy Westland Helicopters.)

A single main rotor helicopter must of course be trimmed in roll, which reduces the thrust potential of the rotor. The ABC™ rotor achieves its roll trim by using two rotors, one balancing the other in roll. The combination of this symmetry with the high stiffness of the rotor hubs and blades allows each rotor to operate out of trim and make full use of the aerodynamic thrust potential of the advancing side in high speed forward flight.

## Blown or circulation control rotors

These make use of the circulation control aerofoil[3] to alleviate the stall problems on the disc. An aerofoil obtains lift by inducing a circulatory flow and when combined with the incident flow causes the pressure difference between the upper and lower surfaces and hence derives lift. This circulation is generated by the incidence angle between the aerofoil and the incoming flow direction. Ultimately this flow will break down and the aerofoil will stall. A circulatory flow can be set up on profiles other than a recognised aerofoil section, including circular, by blowing air from within the body through slots aligned tangentially to the body surface. Indeed, an aerofoil section can generate lift in the reversed flow region by blowing along the upper surface in a direction from the trailing edge to the leading edge (see Fig. 2.1). It is this technique which forms the principle of circulation control. An advantage is that the circulation and hence lift is obtained in a manner not involving an incidence, and is therefore not subject to the same type of flow breakdown and stall. It is used to control the flow over a surface when separation needs to be avoided. This can be to prevent stall or to generate lift in situations which would not be otherwise obtainable. A considerable increase in forward speed or manoeuvre/lift capability can be achieved. The complexities of supplying the blowing pressure along the blade slots will cause weight penalties.

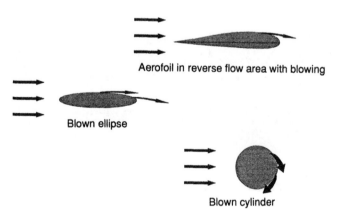

Aerofoil in reverse flow area with blowing

Blown ellipse

Blown cylinder

**Fig. 2.1.** The principle of circulation control.

## Stopped rotor

This concept[4] is the limit of compounding. The aircraft takes off and lands as a helicopter, but during cruise reverts to a fixed wing aircraft by slowing the rotor to a halt and transferring the lifting responsibilities to a wing. During the rotor slowing process the dissymmetry between the advancing and retreating sides of the disc increases and circulation control can be used to offset this effect and enable roll trim to be achieved.

There are obvious weight penalties, because in any flight condition, either the wing or the rotor is unproductive dead weight. At low rotor speeds there can be control problems with the rotor, which is particularly apparent in gusty conditions.

## X wing

This is a combination of the philosophies of the blown and stopped rotors.[5] The rotor blade section is symmetric, blown, and can be aerodynamically reversed by changing the blowing pattern. When the rotor is stopped (at 45° to the fuselage centre line—hence the configuration name) the blowing is applied so that the rotor blades themselves act as the wings. The blowing control is very complicated and computer controlled with the greatest complications arising during transition from and to forward flight from the hover.

**Plate 2.9.** Bell Boeing V22 Osprey tilt rotor aircraft completing initial shipboard compatibility tests aboard the USS *Wasp*, Dec. 1990. (Photo courtesy of Bell Helicopter Textron Inc.)

## Tilt wing/tilt rotor

These designs[6] require the axis of the rotors to be turned through at least a quadrant and thereby be a helicopter or a fixed wing propeller driven aircraft. The rotor design is naturally a compromise between a large aspect ratio/low twist helicopter rotor and a smaller aspect ratio/high twist propeller and will suffer accordingly. However, the importance of these concepts is shown by the continual return of the industry to these configurations, the Osprey being the natural, most recent example. See Plate 2.9.

The transmission will be complicated and interference with the wing will be present. (This will be reduced for the tilt wing aircraft where the thrust line of the rotor is kept parallel to the chordline of the wing.) Transition is the most problematic flight regime.

# 3

# The rotor in hover

## Introduction

Aerodynamically, hover is the most straightforward of the flight regimes to model theoretically. That does not mean that it is easy, but it does possess the distinct advantage of axial symmetry. This chapter introduces two types of approach: the momentum and blade element theories. In the first method, typified by the actuator disc, the realities of rotor blades are ignored and the rotor is treated as a single entity obtaining thrust by setting up a uniform mass flow of air over its whole plan area and imparting a change of momentum to it. This is a simplistic view of the problem but it highlights some important general facts about the performance of a rotor in hover. The second approach is aimed at treating each rotor blade as a wing and introducing the usual lift and drag specifications as used for fixed wing analyses. The variation of airflow speed over the blade according to its radial position is a complication and the vortices springing from the blade tips, in particular, form a helical pattern underneath the rotor disc which has a great influence on the rotor performance. The description of a detailed vortex model is beyond the scope of this book, but the relevant chapters will indicate how simpler methods can be derived.

## Actuator disc theory

The basic characteristics of the rotor in the hover and axial (vertical) flight can be obtained from simple actuator disc theory which is based on a momentum approach.

The assumptions made by this model are as follows:

1.  The rotor has an infinite number of blades (i.e. a totally uniform system).
2.  The rotor is modelled by a constant pressure difference across the disc.
3.  The velocity of the air downward through the disc (downwash), is constant across the disc.
4.  The vertical velocity is continuous through the rotor disc.
5.  There is no swirl in the wake (i.e. no rotational velocity is imparted to the air passing through the disc, all flow velocities are axial).
6.  The airflow is divided into two distinct regions, namely that which passes through the rotor disc and that which is external to the disc. The division of these two types of flow is a streamtube which passes through the perimeter of the disc. The wake forms a uniform jet.

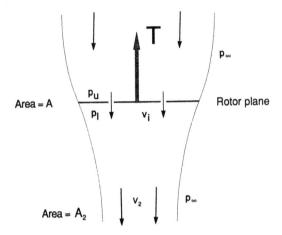

**Fig. 3.1.** Actuator disc streamtube.

The actuator disc model for the hovering rotor is shown in Fig. 3.1.

## Nomenclature

Disc area = $A$
wake area = $A_2$
atmospheric static pressure = $p$
atmospheric density = $\rho$
atmospheric pressure above the rotor disc = $p_u$
atmospheric pressure below the rotor disc = $p_l$
induced velocity induced at the rotor = $v_i$
velocity in the wake, far downstream = $v_2$.

The flow in the wake is assumed to come from air ahead of the rotor at rest with pressure $p_\infty$, and to return to pressure $p_\infty$ far downstream of the rotor.

The rotor thrust is given by

$$T = A(p_l - p_u) \tag{3.1}$$

Applying Bernoulli s equation above the rotor gives

$$p_\infty = p_u + \tfrac{1}{2}\rho v_i^2 \tag{3.2}$$

Applying Bernoulli s equation below the rotor gives

$$p_\infty + \tfrac{1}{2}\rho v_2^2 = p_l + \tfrac{1}{2}\rho v_i^2 \tag{3.3}$$

From Eqns (3.1)–(3.3) we have

$$T = \tfrac{1}{2}\rho v_2^2 A \tag{3.4}$$

In addition, the rotor thrust is equal to the rate of change of axial momentum of the air per unit time. To evaluate this, the mass flow through the rotor disc per unit time is

$$M_{\text{FLOW}} = \rho A v_i \tag{3.5}$$

The thrust can now be expressed by

$$T = M_{\text{FLOW}} v_2 = \rho A v_i v_2 \tag{3.6}$$

In every second, air of mass $M_{\text{FLOW}}$ enters the streamtube with zero vertical velocity, whilst an equal mass of air leaves the streamtube with vertical velocity $v_2$. In this time period the air in the streamtube acquires a vertical momentum of $M_{\text{FLOW}} v_2$.

From (3.4) and (3.6) we find that

$$v_2 = 2v_i \tag{3.7}$$

From which substituting into (3.4) and rearranging gives

$$v_i = \sqrt{\frac{T}{2\rho A}} \tag{3.8}$$

The value for $v_i$, so obtained, is called the ideal induced velocity and, because it is uniform over the rotor disc, is the minimum value for a given rotor thrust. The importance of this result lies in the predictions it makes for the power required to generate this rotor thrust. In hover, the greatest source of power is that required to maintain the rotor thrust $T$ working against an air inflow of $v_i$. This is called the thrust induced power (or induced power for brevity) and is given by

$$P_i = T v_i \tag{3.9}$$

Or using (3.8) by

$$P_i = T \sqrt{\frac{T}{2\rho A}} \tag{3.10}$$

and is called the *ideal induced power*.

An indication of the hover performance efficiency can be derived from the above analysis using, as a measure, the thrust per unit power required, i.e.

$$\frac{T}{P_i} = \frac{1}{v_i} \tag{3.11}$$

**Table 3.1**

| Aerospace vehicle designation | Aerospace vehicle type | Equivalent disc loading (N/m²) | Ideal induced velocity (m/s) | Hover efficiency (N/kW) |
|---|---|---|---|---|
| Mil Mi-10k | Crane helicopter | 388 | 12.6 | 79.4 |
| Westland Lynx | Utility helicopter | 364 | 12.2 | 82.0 |
| Sikorsky S64A | Crane helicopter | 496 | 14.2 | 70.3 |
| Sikorsky CH-53E | Heavy helicopter | 718 | 17.1 | 58.5 |
| Bell V22 Osprey | Tilt rotor | 1161 | 21.8 | 45.9 |
| Hiller-Ryan XC-142 | Tilt wing | 2387 | 31.2 | 32.1 |
| Ryan XV5A | Fan lift | 16823 | 82.8 | 12.1 |
| Harrier GR3 | Jet lift | 114403 | 216 | 4.6 |
| Space shuttle | Rocket | 465479 | 436 | 2.3 |
| Saturn V F1 | Rocket | 2964912 | 1087 | 0.9 |

Typical values for this parameter, for a selection of thrusting devices, are given in Table 3.1 and shown in Fig. 3.2.

It is immediately apparent that a lower disc loading carries the benefit of a better (more efficient) hover performance. The induced power is not the only power drain on the engines, but in the hover it accounts for a high proportion of the total (in the region of 70%) and in consequence is extremely influential on the overall hovering performance of a helicopter.

## Thrust and power coefficients

In order to compare different rotor designs in a sensible manner, thrust and power coefficients are used. These quantities mirror the lift and drag coefficients used for a fixed wing aircraft.

For a fixed wing aircraft a lift or drag coefficient is based on the forward flight speed and the wing area. For the helicopter rotor, however, these are not applicable, and in this case the reference speed is taken to be the rotor tip speed (since zero forward speed in the hover renders an infinite coefficient), and the reference area is that of the complete rotor disc. A variant for the reference area is that of the total blade area. This is introduced in due course.

The thrust coefficient is defined by

$$C_T = \frac{T}{\frac{1}{2}\rho V_T^2 A} \tag{3.12}$$

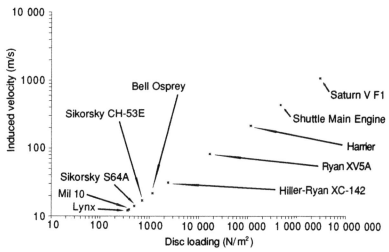

**Fig. 3.2.** Hovering efficiency for various vehicles.

The power coefficient is defined by

$$C_P = \frac{P}{\frac{1}{2}\rho V_T^3 A} \tag{3.13}$$

Here $V_T$ is used to denote the rotor tip speed, and $A$ is the rotor disc area. It is important to note the denominator contains the factor of a half. Some references will not include the half and this difference is important to watch for and consultation of the relevant basic text glossary is recommended.

Using the above analysis the ideal downwash, thrust and induced power coefficients are connected by the following expressions

$$\lambda_i = \tfrac{1}{2}C_T^{1/2}$$
$$C_{Pi} = \tfrac{1}{2}C_T^{3/2} \tag{3.14}$$
$$= C_T \lambda_i$$

Typical values for $C_T$ are in the range 0.01–0.05.

## Non-ideal behaviour of the rotor

The characteristics of a real rotor differ from the ideal situation which has just been described largely through two effects.

### Viscous effects

Viscous effects on a wing give rise to profile drag and hence, to overcome this, an extra demand is made, known as profile power, which is that required to turn the rotor against the viscous drag forces on the blades. It is present irrespective of whether the rotor is producing a thrust or not and its magnitude is normally below the other power drains on the rotor. For instance, in hover the induced power dominates, however, when the rotor blades penetrate the stall boundary, the profile drag forces on the blades increase substantially with a consequent increase in profile power.

### Induced velocity effects

The momentum theory assumes uniformity over the entire rotor disc, particularly the air flow velocity through it. However, in reality, this induced velocity varies across the disc and these differences can arise from various sources: tip effects, loading effects, swirl effects and the ground effect, each of which will be considered in this chapter.

## Tip effects

For obvious reasons, the blade loading must fall to zero at the tip. In fact a region of the blade just inboard of the tip will not be producing lifting forces. To produce lift, a pressure difference must be generated by the rotor blade, however, at the blade tip, the air can by-pass the pressure difference by "sneaking" around the outside edge of the blade tip. This reduction of the effective lifting potential of the rotor blade tips requires special treatment, which, in its simplest form, is to specify the extent of the blade tip region which is effectively producing no lift and regards the outermost region of the blade as missing for any analysis of the blade lift. The region is defined by a tip loss factor which is usually expressed as a fraction of the blade radius. The tip loss factor is usually denoted by $B$, making the effective rotor radius as $BR$.

There are various models used to define the tip loss factor, three of which are:

(a) based on rotor loading

$$B = 1 - \sqrt{\tfrac{1}{2} C_T} / N \quad \text{(Glauert)}[7]$$

(b) Based on blade geometry

$$B = 1 - \text{Chord}_{\text{TIP}} / 2R$$

(c) From the previous model using typical conventional planform data and justified by flight test:[8]

$$B = 0.97, \text{ where}$$
$$N \text{ is the number of blades.}$$

## Loading effects

The actuator disc model which has already been described assumes that the downwash is constant across the rotor disc. However, in reality the downwash is not uniform along a rotor blade due to various factors such as aerofoil section and blade chord. A constant downwash distribution has benefits as regards efficiency so any means of achieving this will have important effects on the hovering performance of a helicopter. Because the incident air speed parallel to a blade chord varies along the span of a rotor blade, the loading potential is greater in the tip region.

Additionally, the blade will leave a trailing vortex system which will roll up under its own influence and like a fixed wing wake will tend to form a concentrated vortex near the tip. In the case of a helicopter blade this rolled-up vortex will position itself inboard of the blade tip and because of its proximity to the blade will induce a considerable vertical velocity on the tip flow. Indeed, in some circumstances, this can be sufficient to give a net upwash.

## *Prescribed wake models*

Prescribed wake models actually define the position in space of the vortices comprising the rotor wake and are often the result of experimental investigations.

Figures 3.3 and 3.4 show an example of such a method[9] developed to predict the position of vortices in the rotor wake or slipstream. Figure 3.3 shows the radial and vertical position of the vortices against rotor rotation and Fig. 3.4 their position in space. It should be emphasised that the initial downward movement of the vortex rapidly increases as the following vortex is shed. The initial motion is radially inward and an abrupt downward movement from then on. This places a vortex close to the following blade causing important interference effects. Indeed, some modern blade designs have the tip area of the blade structurally cranked downwards to move this vortex away from the blade with the aim of reducing its interference and improving the hovering efficiency of the rotor.

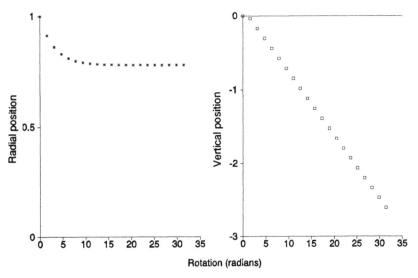

**Fig. 3.3.** Vortex model, vortex positions with respect to rotor rotation.

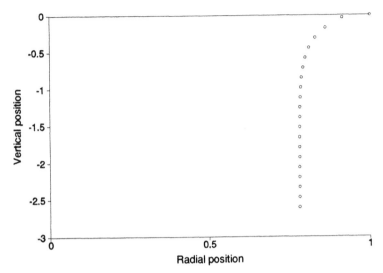

**Fig. 3.4.** Vortex model, vortex position in space.

## Free wake models

A free wake model, in contrast, allows the rotor wake to develop from nothing, under its own influence. Each blade tip vortex will influence itself as well as other vortices and so as the wake develops and more vortices need to be positioned, the amount of calculation increases dramatically (roughly as the square of the number of vortices) and so considerable computing power is usually required for all but the smallest of wakes. The method must allow for the ageing process of the vortex as viscous action and instability distort its character. The vortex filament, which is so useful for many applications, has the danger of an infinite induced velocity for a point actually lying on it. So as a free wake develops, any vortex filaments coming closely together will generate large velocities and the wake will develop with the danger of violent vortex movements. This requires a tight control of the calculations.

So the constant loading which an actuator disc assumes, does not occur in practice. The ideal situation will not be the case and for a given thrust the induced power will be in excess of that predicted by the momentum theory. (This is analogous to the minimum induced drag of an elliptically loaded wing which gives a uniform downwash distribution across the span.)

In order to examine the effect of a non-uniform downwash distribution the idea of an annulus theory needs to be described. Rather than the entire rotor disc being used for the momentum analysis, the disc is divided into a series of annuli, which are infinitesimally small in width, and then apply to each annulus the analysis which has previously been described. The overall rotor performance is then determined by integrating the results for a general annulus over the complete rotor.

The general annulus has radius $r$, width $dr$ and a local downwash $v_i(r)$. The annulus area is then

$$2\pi r\, dr \tag{3.15}$$

then the mass flow through the annulus is

$$2\pi \rho r\, dr\, v_i \tag{3.16}$$

As before, the vertical velocity far downstream is given by $2v_i$, therefore the thrust obtained from the annulus is equal to the rate of change of momentum of the flow passing through it, i.e.

$$dT = 2\pi \rho r\, dr\, v_i\, 2v_i = 4\pi \rho v_i^2 r\, dr \tag{3.17}$$

The induced power for the annulus is then

$$dP_i = 2\pi \rho r\, dr\, v_i\, 2v_i\, v_i = 4\rho \pi v_i^3 r\, dr \tag{3.18}$$

Upon integrating over the whole rotor disc we have

$$T = 4\pi \rho \int_0^R v_i^2 r\, dr \tag{3.19}$$

$$P_i = 4\pi \rho \int_0^R v_i^3 r\, dr \tag{3.20}$$

which are general results applicable for any induced velocity distribution, $v_i(r)$. As an example, if we now assume that the induced velocity distribution is expressible as a power of rotor radius, we have the following result

$$v_i(r) = v_i^T \left( \frac{r}{R} \right)^n \tag{3.21}$$

where zero downwash flow is assumed at the rotor centre and $v_i^T$ is the value of the downwash at the blade tip.

Using the above results we find, by substitution, that the rotor thrust is given by

$$T = \frac{4\pi \rho v_i^{T2}}{R^{2n}} \int_0^R r^{(2n+1)}\, dr = \frac{4\rho A v_i^{T2}}{2n+2} \tag{3.22}$$

Similarly the induced power is given by

$$P_i = \frac{4\pi \rho v_i^{T3}}{R^{3n}} \int_0^R r^{(3n+1)}\, dr = \frac{4\rho A v_i^{T3}}{3n+2} \tag{3.23}$$

Now from the thrust expression, (3.22), we have

$$v_i^T = \sqrt{n+1}\sqrt{\frac{T}{2\rho A}} \tag{3.24}$$

and noting that the ideal induced power is given by

$$P_{i\,\text{IDEAL}} = T\sqrt{\frac{T}{2\rho A}} \tag{3.25}$$

we find that the expression for the induced power, (3.23), becomes

$$P_i = \left(\frac{2n+2}{3n+2}\right)T\sqrt{n+1}\sqrt{\frac{T}{2\rho A}} \tag{3.26}$$

whence

$$\frac{P_i}{P_{i\,\text{IDEAL}}} = \left(\frac{2n+2}{3n+2}\right)\sqrt{n+1} = \frac{(n+1)^{3/2}}{1+\dfrac{3n}{2}} = k_i \tag{3.27}$$

where $P_{\text{IDEAL}}$ is the ideal induced power of the actuator disc given by Eqn (3.10).

The term $k_i$, as defined above, is the amount that the induced power varies compared with the ideal case of uniform downwash ($n = 0$) and expressed as a ratio.

The effect of various values of $n$ on $k_i$ are

| $n$ | 0 | 0.5 | 1 | 2 | 3 | 5 |
|-----|-----|------|------|------|------|------|
| $k_i$ | 1.0 | 1.05 | 1.13 | 1.30 | 1.45 | 1.73 |

It can thus be seen that as the downwash becomes more peaky towards the tip, increasing the value of $n$, so the induced power factor $k_i$ increases.

## Swirl effects

The spiral nature of the vortex wake means that a swirl velocity will be induced in the airflow as it passes through the rotor which has the effect of causing the wake to rotate in the same direction as the rotor. The effective rotor speed is then reduced and the useful dynamic head is consequently lower. This will naturally cause a reduction in the lift efficiency of the rotor and therefore the induced power for a given rotor thrust will be increased.

# Ground effect

This is a well known effect for fixed wing aircraft, which is also called the ground cushion, where an aircraft close to the ground tends to float. This can be viewed as a potential increase in lift capacity or a reduction in power for the same lift. In either case the aircraft is effectively flying in an upwash created by a mirror image of itself relative to the ground.

The image ensures that the flow directions at ground level are horizontal and the vertical component zero. The benefits of operating in an upwash are used by birds in flight by using the often observed V formation. The leading bird creates tip vortices which generate upwash on either side. The birds immediately behind and outside of the leading bird will therefore benefit. This benefit is added to by the succeeding birds in the formation and therefore helping the following birds.

The helicopter, when close to the ground, also sees an image of itself, as shown in Fig. 3.5 and will therefore benefit from the virtual upwash. The benefits are again seen as an increase in lift for a given power, or a power reduction for the same lift. The effect has been modelled by various methods such as sources or vortices.[10] However, it is important to emphasise that specifications of helicopter performance in the hover are usually quoted twice, i.e. OGE (out of ground effect) or IGE (in ground effect) and the beneficial presence of the ground in creating this virtual upwash is the reason for the difference. The presence of buildings close to

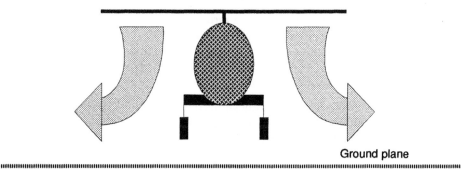

Ground plane

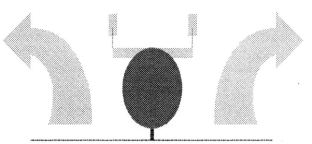

**Fig. 3.5.** Ground effect—helicopter and its image in the ground plane.

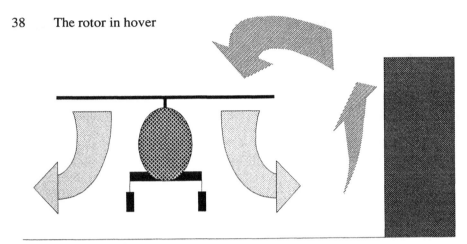

**Fig. 3.6.** Influence of buildings on a hovering helicopter.

a hovering helicopter can reduce or indeed reverse the benefits of ground effect by causing the airflow spilled along the ground to be turned upwards and eventually to pass through the rotor again. This will create a recirculation through the rotor which degrades its performance (see Fig. 3.6).

## Blade element theories in the hover

Actuator disc theory provides a good insight into some important facts about the helicopter rotor but is limited in its scope. The following questions need a more sophisticated means of analysis to be answered:

1.  What blade pitch is required to give a thrust, *T*?
2.  Can the ideal of a uniform downwash be achieved?
3.  What is the power required to overcome the profile drag of the rotor blades?
4.  How can a rotor design be optimised with a set of given constraints on the rotor parameters such as tip speed and blade geometry?

In order to answer these and other questions the theory must be capable of modelling the blade geometry and the aerofoil performance correctly. The mechanism is to use blade element theory, which can be combined with a downwash model, like actuator disc theory, to give the more extensive, realistic and complete examination of a helicopter rotor. This comment is not to condemn momentum theory but to emphasise its limitations.

We will again restrict the discussion to hover, but the theory can be extended to axial flight with no difficulty.

Figure 3.7 shows the flow direction incident on a typical aerofoil section at radius *r* from the rotor centre. To continue the analysis, two aspects of a rotor need to be highlighted:

1. The collective pitch angle ($\theta_0$) is the pitch angle applied to the rotor blade and remaining constant as it rotates around the rotor shaft. There is another pitch angle (cyclic), which will be discussed later, in which changes are made to the blade as it rotates around the azimuth.

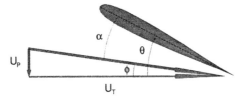

**Fig. 3.7.** Incident flow direction.

2. The dynamic pressure of the air varies as the square of the rotor radius. If the lift coefficient distribution along the blade, determined by the local pitch angle, remains essentially constant the resulting blade loading will be heavily biased towards the blade tip and the downwash distribution will be very peaky in the tip region. As already shown, this will give rise to a high value of induced power factor, $k_i$, degrading the hovering efficiency of the rotor. Because of this, a main rotor blade is usually twisted along its length in order to even out the lift distribution (see Plate 3.1). This is to avoid a concentration of lift at the tips where there may be structural problems and as explained will move the rotor away from a uniform downwash condition and its attendant power benefits. A tail rotor blade is usually untwisted because of its very different aerodynamic requirements when

**Plate 3.1.** Rotor blade of a Boeing Vertol Chinook tandem rotor helicopter showing the built-in blade twist.

compared to the main rotor. This is discussed in Chapter 11. The twist is usually expressed as an overall angle change along the complete rotor blade and its projection, from the hub centre to the blade tip. Although the blade is attached to the rotor hub at a point outboard of the rotor hub centre line, the twist is imagined to continue to the rotor centre and the resulting increased figure used for the twist. It is usual to express the twist by the term $\kappa$ which is often constant giving a linear variation of twist with radial position. Modern rotors have non-linear twist variations along the blades.

For this first look at blade element theory, a linear twist distribution is assumed and the downwash $v_i$ is assumed constant over the whole rotor disc.

At the general radial position $r$, the pitch angle can be expressed by

$$\theta = \theta_0 - \kappa \frac{r}{R} \tag{3.28}$$

The right-hand side consists of the collective pitch minus the linear twist.

The inflow speed components, $U_P$ and $U_T$, are given by

$$U_T = \Omega r \quad \text{and} \quad U_p = v_i \tag{3.29}$$

For the inflow angle generated by the downwash we have

$$\phi = \tan^{-1}\left(\frac{U_P}{U_T}\right)$$
$$= \tan^{-1}\left(\frac{v_i}{\Omega r}\right) \tag{3.30}$$

The angle of incidence follows

$$\alpha = \theta - \phi = \theta_0 - \kappa\left(\frac{r}{R}\right) - \frac{v_i}{\Omega r} \tag{3.31}$$

where $\phi$ has been assumed to be small.

The lift over a small chordwise strip of the blade, of width $dr$, at a general rotor blade station $r$, is given by

$$dL = \frac{1}{2}\rho(\Omega r)^2 c\,dr\,a\left[\theta_0 - \kappa\left(\frac{r}{R}\right) - \frac{v_i}{\Omega r}\right] \tag{3.32}$$

As the inflow angle, $\phi$, is small, we have for the $N$ blades

$$T = N\int_0^R dL = \frac{1}{2}\rho\Omega^2 caN\int_0^R\left[\theta_0 - \kappa\left(\frac{r}{R}\right) - \frac{v_i}{\Omega r}\right]r^2\,dr \tag{3.33}$$

It is usual to non-dimensionalise the rotor geometry terms by the radius $R$ and any flow velocities by the rotor tip speed $V_T$. Using the substitutions $V_T = \Omega R$ and $r = Rx$ we find

$$
\begin{aligned}
T &= \frac{1}{2}\rho V_T^2 NcRa \int_0^1 \left[ \theta_0 - \kappa x - \frac{\lambda_i}{x} \right] x^2 \, dx \\
&= \frac{1}{2}\rho V_T^2 NcRa \left[ \frac{\theta_0}{3} - \frac{\kappa}{4} - \frac{\lambda_i}{2} \right]
\end{aligned}
\tag{3.34}
$$

The solidity of a rotor, denoted by $s$, is defined by

$$
s = \frac{\text{blade area}}{\text{disc area}} = \frac{NcR}{A} = \frac{NcR}{\pi R^2} = \frac{Nc}{\pi R}
\tag{3.35}
$$

The blade element theory, because of its approach, must necessarily depend on the blade area of the rotor ($NcR$). The thrust coefficient, defined in (3.12), comes from a momentum analysis which focuses on the entire rotor as a lifting device and therefore is only concerned with the total disc area $A$. As this analysis is combining results from the two different theories, the ratio of these areas is a natural result, and this is the rotor solidity. Therefore

$$
\frac{C_T}{s} = \frac{T}{\frac{1}{2}\rho V_T^2 NcR}
\tag{3.36}
$$

while the thrust coefficient $C_T$ is based on disc area, it can be seen that the $C_T/s$ term has the same form but is based on the blade area.

Therefore (3.34) becomes

$$
\frac{C_T}{sa} = \frac{\theta_0}{3} - \frac{\kappa}{4} - \frac{\lambda_i}{2}
\tag{3.37}
$$

Now from momentum (actuator disc) theory (3.14) we have

$$
C_T = 4\lambda_i^2
\tag{3.38}
$$

and substituting the result in (3.37) gives

$$
\lambda_i^2 + \frac{sa}{8}\lambda_i - \frac{sa}{12}\left( \theta_0 - \frac{3}{4}\kappa \right) = 0
\tag{3.39}
$$

Note that $\theta_0 - 3\kappa/4$ is the geometrical pitch angle at 75% rotor radius. This analysis allows a short cut calculation method to be derived, which is based on a comparison of the twisted rotor blade result as presented in (3.39) with that of an equivalent untwisted rotor blade. If the downwash as predicted by (3.39) is equal to that obtained by an untwisted blade set at a representative collective pitch $\bar{\theta}_0$

then (3.39) for the untwisted blade becomes

$$\lambda_i^2 + \frac{sa}{8}\lambda_i - \frac{sa}{12}(\bar{\theta}_0) = 0 \qquad (3.40)$$

from which we find $\bar{\theta}_0 = \theta_0 - 3\kappa/4$ and is the pitch angle at the 75% radius. That is, the correct downwash is obtained by taking the pitch angle at the 75% radius and assuming this applies along the whole blade.

Hence for "back of the envelope" quick calculations for a rotor where a representative aerodynamic station is required to model the whole blade, the 75% radial station is very useful. For a more general (non-linear) twist distribution, the twist term in (3.28) will have to be included explicitly.

To illustrate the method, an example is presented. The rotor data are as follows:

| | |
|---|---|
| solidity ($s$) | 0.078 |
| lift curve slope ($a$) | 5.8 per radian |
| induced power factor ($k_i$) | 1.1 |
| profile drag coefficient ($C_{D0}$) | 0.011 |
| blade twist | 8° |

Figures 3.8–3.10 show the variation with collective pitch of downwash, thrust and power coefficients. Figure 3.11 shows the radial distribution of incidence for the three collective pitch angles 6°, 12° and 18°. The case of 6° collective pitch gives zero thrust, and hence downwash. The incidence for this case is purely the geometric pitch angle.

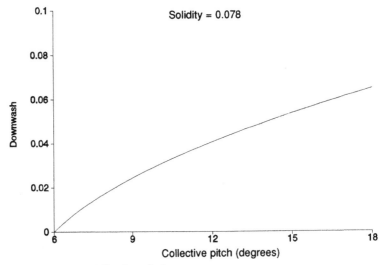

**Fig. 3.8.** Downwash vs collective pitch.

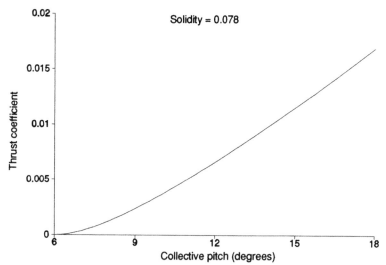

**Fig. 3.9.** Thrust coefficient vs collective pitch.

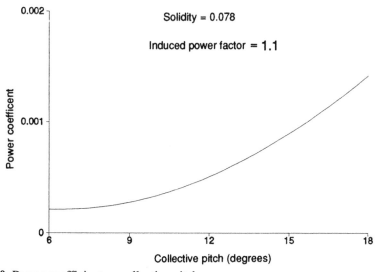

**Fig. 3.10.** Power coefficient vs collective pitch.

# Annulus theory

The combination of momentum and blade element theories, as previously described, assumed a constant downwash over the whole rotor disc. To examine the effect of a non-uniform flow velocity through the rotor, the method is extended to evaluate the rotor, allowing for a variation of downwash along the blade span.

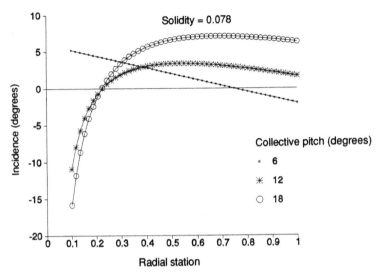

**Fig. 3.11.** Radial incidence distribution.

For this analysis, the following assumptions are made:

1. Because rotor blades have high aspect ratio, each section or strip can be regarded as behaving in a two-dimensional way independent of the other sections. Two-dimensional aerodynamics can then be used.
2. The rotor disc can be divided into a series of annuli. Each may be regarded as an actuator disc in order to relate the local induced velocity to the lift developed by the blade sections sweeping out at that particular radius.

Consider a blade section at radius $r$ from the centre of rotation, as shown in Fig. 3.12, where $v_i(r)$ is the induced velocity, $dL$ is the section lift, $dD$ is the section drag, $dT$ is the section thrust and $d\hat{H}$ is the section torque. In the analysis, $v_i$ will be assumed to be small compared with $\Omega r$. This assumption is reasonable for the majority of the rotor disc.

The thrust developed by the $N$ blade segments in the annulus between $r$ and $r + dr$ is given by

$$N\,dT \simeq N\,dL = N\frac{1}{2}\rho(\Omega r)^2 a\left(\theta - \frac{v_i}{\Omega r}\right)c\,dr \tag{3.41}$$

where $a$ is the section lift curve slope and $c$ is the local blade chord at radius $r$.

Application of actuator disc theory to this annulus, using Eqn (3.8), where the thrust produced by the annulus of area $2\pi r\,dr$ is $N\,dT$ gives

$$v_i(r) = \sqrt{\frac{N\,dT}{2\rho 2\pi r\,dr}} \tag{3.42}$$

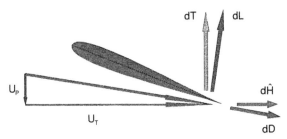

**Fig. 3.12.** Forces on blade element.

from which

$$N \, dT = 4\pi\rho r \, dr \, v_i^2(r) \tag{3.43}$$

Eliminating $N \, dT$ between (3.41) and (3.43) and rearranging gives

$$v_i^2(r) = \frac{1}{8} V_T^2 \left( \frac{c(r)}{\bar{c}} \right) sa \left( \theta \frac{r}{R} - \frac{v_i(r)}{V_T} \right) \tag{3.44}$$

where $s$ is the *solidity* of the rotor, defined using $\bar{c}$ the mean blade chord. For rotor blades of rectangular planform, $c = \bar{c}$ (i.e. constant).

Points to note from Eqn (3.44) are that it is a quadratic equation which relates the local induced velocity $v_i(r)$ to the local pitch angle $\theta$ for a given rotor and that blade section characteristics such as $c(r)$ and $a(r)$ are now involved.

## Ideal twist

One possible way of making $v_i(r)$ uniform over the disc (i.e. independent of $r$), and in consequence giving the minimum induced power for a given rotor thrust, requires the following

$$c = \text{constant}$$
$$a = \text{constant}$$
$$\text{and } \theta * (r/R) = \text{constant}$$

i.e. $\theta$ must be proportional to $1/r$ resulting in what is termed *ideal twist*. As $r$ tends to zero, the twist tends to infinity with an obvious problem in interpretation but this will not happen in reality as the blades have a root cut-out. This is the radius at which the aerofoil section begins. It is usually expressed as a fraction of the rotor radius $R$. The main limitation to the concept of ideal twist is that it applies to a single flight state of the rotor only. In other words, once a blade is manufactured, the twist will be optimum for only one flight condition. Usually, for ease of manufacture, rotor blades were made with a linear twist over the entire blade, a value of 8° is typical. As already mentioned, tail rotor blades have twists which differ from that of the main rotor, and are quite often zero. The advent of aerofoil sections designed to cater specifically for helicopter rotors, has meant that aerofoil sections can change along the blade span. This means that changes in the no-lift angle along the span will affect the twist in addition to the geometric shape and, as an example, on the EH101 main rotor a non-linear twist is very evident, especially when viewed from above. Modern composite manufacturing techniques have made the introduction of sophisticated blade profiles possible.

The total rotor thrust $T$ is obtained by integrating Eqn (3.43)

$$
\begin{aligned}
T &= \int_{\text{BLADE}} N \, dT \\
&= \int_0^R 4\pi \rho r v_i^2(r) \, dr
\end{aligned}
\tag{3.45}
$$

The rotor power is given by

$$
\begin{aligned}
P &= \int_{\text{BLADE}} \Omega r N \, d\hat{H} \\
&= \int \Omega r N \, (dD + dL \cdot \phi) \\
&= \int \Omega r N \left( dD + dL \cdot \frac{v_i}{\Omega r} \right) \\
&= \int \Omega r N \, dD + \int \Omega r N \, dL \frac{v_i}{\Omega r} \\
&= \int_0^R \Omega r N \, dD + \int_0^R v_i N \, dL
\end{aligned}
\tag{3.46}
$$

The first term is called the *profile power* because it arises from the blade profile drag. The second term is the *induced power* already discussed.

Now

$$
dD = \frac{1}{2} \rho (\Omega r)^2 C_D c \, dr
\tag{3.47}
$$

where $C_D$ is the aerofoil section profile drag coefficient. On substituting this expression for $dD$ in Eqn (3.46) and integrating, the first term in the rotor power expression (3.46) becomes, with the assumption of constant chord

$$
\frac{1}{8} \rho N c C_D \Omega^3 R^4
\tag{3.48}
$$

which is termed the profile power. The integrand contained in the second term of (3.46) is $v_i \cdot N \, dL$ and is the product of the thrust produced over an annulus together with the local value of induced velocity. Earlier in this chapter, Eqn (3.9) showed a similar result except that it was for the entire rotor. This must be the case since (3.9) is a result of using an actuator disc approach to the entire rotor which necessitates the assumption of constant induced velocity, $v_0$, over the rotor disc. If the $v_i$ term in (3.46) is assumed constant and set to $v_0$, then the power becomes

$$P = \frac{1}{8}\rho N c C_D \Omega^3 R^4 + T v_0 \tag{3.49}$$

Recalling (3.8), this can be expressed as

$$P = \frac{1}{8}\rho V_T^3 N c R C_D + \frac{T^{3/2}}{\sqrt{2\rho A}} \tag{3.50}$$

From the above analysis it can be seen that the induced power is controlled by the disc area whilst the profile power by the blade area, tip speed $V_T$, or profile drag coefficient $C_D$. The total power will depend on both, and the rotor design must reflect this. Usually in or near to the hover the total power will be dominated by the induced component and so efficient operation in this flight condition will determine the disc area and hence the main rotor radius. In forward flight, especially if the high speed end of the flight envelope is being examined, the induced power dies away, leaving the profile power when, as can be seen from Eqn (3.50), the blade area becomes the most important term. The blade chord is then determined by these flight conditions, as well as giving the main rotor sufficient blade area to achieve the required rotor thrust for the high speed end of the flight envelope. Other design aspects will need to be addressed and will be discussed in later chapters.

## Figure of merit

This quantity is a means of expressing the power actually consumed by a rotor when compared to the ideal minimum of an induced power factor of unity (actuator disc), and no profile drag $(C_{D0} = 0)$. The ideal induced power situation must be regarded as perfection and the figure of merit gauges the hovering performance of an actual rotor with the theoretical optimum.

The definition is

$$\text{figure of merit} = \frac{\text{ideal induced power}}{\text{actual power}} \tag{3.51}$$

The figure of merit value will vary as the rotor thrust increases and the induced power becomes more dominant. However, it serves as a quick method of gauging the performance of a rotor in hover and takes values, typically, of the order of $0.6 - 0.75$. Figure 3.13 shows the variation of figure of merit for the rotor used in the blade element theory examples shown in Figs 3.8–3.11.

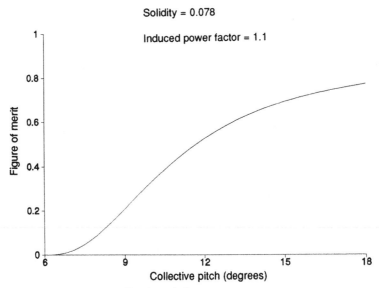

**Fig. 3.13.** Figure of merit vs collective pitch.

## Ducted fans

With suitable design, the performance of a propeller or rotor is improved by placing it within a duct or shroud. This application has been used in some helicopter
designs, examples being the fenestron tail rotor installation on the SA 341 Gazelle
or the lifting fans of the Bell X22A. The idea of shrouding is to increase the thrust
for a given power. For a shrouded rotor of identical dimensions and power consumption to its unshrouded equivalent, a thrust increase of 1.2 – 1.35 can be
achieved. There is a price for this since the existence of the duct can cause flow
interruption and breakdown in forward flight which can negate the benefits of the
shroud.

A discussion of ducting the rotor is included in Chapter 4.

# 4

# The rotor in axial flight—climb and descent

## Vertical flight performance

The aerodynamics of the hovering rotor can be extended to include axial flight (flight that is parallel to the rotor shaft) by adding the axial velocity to the analyses conducted in the previous chapter. In axial flight, which encompasses vertical climb or descent, axial symmetry is preserved and the assumptions used in the momentum theory can still be applied. This remark has a dangerous side because in some flight conditions, particularly moderate descent speeds, the flow patterns seen in real experiments are dominated by vortex flows which do not exist in the concept of the actuator disc. Hence the extension of the previous theory can be pursued but care is necessary since in certain circumstances the solutions, which exist mathematically, do not conform to the physics of the situation and must therefore be rejected.

One further comment is necessary on the modelling of climb and descent. With hover, air in the streamtube is accelerated into the disc and further accelerated downstream into the wake giving the characteristic contraction. In climb this is unchanged except for the ram effect of the air flowing in towards the rotor disc. The thrust is gained by increasing the downward momentum of the air. In descent, however, air flows upwards into the disc, is slowed as it passes into the disc and is slowed further as it passes above the disc into the wake. The net induced flows are still downward giving the upward thrust direction and the slowing of the air upward through the streamtube preserves its shape.

### The effect of axial translation

The effect of axial translation is to superimpose a climb velocity $V_C$ or a descent velocity $V_D$ on the rotor in hover as shown in Figs 4.1 and 4.2. The momentum, or actuator disc, theory can be modified accordingly to give the induced velocity as follows.

*Climb.* Bernoulli ahead of the rotor

$$p_\infty + \tfrac{1}{2}\rho V_C^2 = p_u + \tfrac{1}{2}\rho(V_C + v_i)^2 \tag{4.1}$$

Bernoulli behind the rotor

$$p_l + \tfrac{1}{2}\rho(V_C + v_i)^2 = p_\infty + \tfrac{1}{2}\rho(V_C + v_2)^2 \tag{4.2}$$

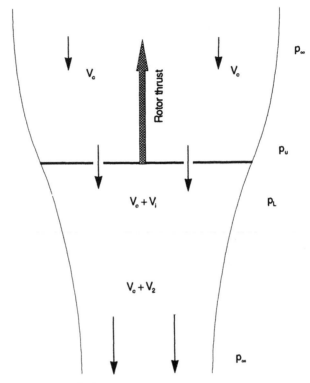

**Fig. 4.1.** The actuator disc in climb.

subtracting (4.1) and (4.2) gives the pressure difference at the rotor disc as

$$p_l - p_u = \rho(V_C v_2 + \tfrac{1}{2} v_2^2) \qquad (4.3)$$

from which the rotor thrust can be expressed by

$$T = A(p_l - p_u) \qquad (4.4)$$

The momentum equation links the thrust with the rate of change of vertical momentum giving

$$T = \rho A(V_C + v_i) v_2 \qquad (4.5)$$

from Eqns (4.3)–(4.5), as in the hover analysis, we find

$$v_i = \tfrac{1}{2} v_2 \qquad (4.6)$$

from which

$$v_i^2 + V_C v_i - \frac{T}{2\rho A} = 0 \qquad (4.7)$$

or

$$v_i^2 + V_C v_i - v_0^2 = 0 \qquad (4.8)$$

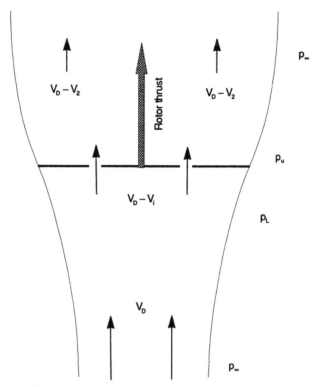

**Fig. 4.2.** The actuator disc in descent.

where $v_0$ is the induced velocity of the rotor in hover at the same thrust value, and is given by

$$v_0 = \sqrt{\frac{T}{2\rho A}} \qquad (4.9)$$

*Descent.* For descent, the analysis now becomes:

Bernoulli ahead of the rotor

$$p_\infty + \tfrac{1}{2}\rho V_D^2 = p_l + \tfrac{1}{2}\rho(V_D - v_i)^2 \qquad (4.10)$$

Bernoulli behind the rotor

$$p_u + \tfrac{1}{2}\rho(V_D - v_i)^2 = p_\infty + \tfrac{1}{2}\rho(V_D - v_2)^2 \qquad (4.11)$$

Subtracting for the pressure difference across the rotor disc gives

$$p_l - p_u = \rho(V_D v_2 - \tfrac{1}{2}v_2^2) \qquad (4.12)$$

The momentum equation gives the thrust as

$$T = \rho A (V_D - v_i) v_2 \tag{4.13}$$

From Eqns (4.4), (4.12) and (4.13) we again have

$$v_i = \tfrac{1}{2} v_2 \tag{4.14}$$

and

$$v_i^2 - V_D v_i + v_0^2 = 0 \tag{4.15}$$

In order to examine the performance of a rotor in axial flight we need to view it in both climb and descent conditions. Equations (4.8) and (4.15) show the induced velocity but with the $V_C$ and $V_D$ terms, assumed positive in the appropriate directions. To compare the results in climb and descent, the substitution $V_D = -V_C$ is made in Eqn (4.15) which becomes

$$v_i^2 + V_C v_i + v_0^2 = 0 \tag{4.16}$$

We can now use Eqns (4.8) and (4.16) to examine the variation of induced velocity with axial velocity.

In order to remove the effect of thrust variation of the rotor, these two equations can be non-dimensionalised by dividing by $v_0^2$ giving the following pair of quadratic equations

$$(\text{climb}) \quad \bar{v}_i^2 + \bar{V}_C \bar{v}_i - 1 = 0, \quad \bar{V}_C \geqslant 0 \tag{4.17}$$

$$(\text{descent}) \quad \bar{v}_i^2 + \bar{V}_C \bar{v}_i + 1 = 0, \quad \bar{V}_C < 0 \tag{4.18}$$

where we have the definitions

$$\bar{v}_i = \frac{v_i}{v_0}, \qquad \bar{V}_C = \frac{V_C}{v_0} \tag{4.19}$$

The solution of these two quadratic equations gives rise to four branches of the induced velocity variation, which are shown in Fig. 4.3. Some parts of the curves exist mathematically, but do not satisfy the physical aspects of the problem and must therefore be rejected. The curves are marked (a)–(e) and are discussed below.

Equation (4.17) gives rise to (a)–(c) while (4.18) gives (d) and (e).

(a) This curve lies wholly below the abscissa and represents a negative value for $v_i$ which cannot occur with the assumed positive thrust. This is physically unjustified and is therefore rejected.
(b) This curve gives a wholly positive solution but, due to factors which are discussed next, is not appropriate for all axial flight conditions. This is only the

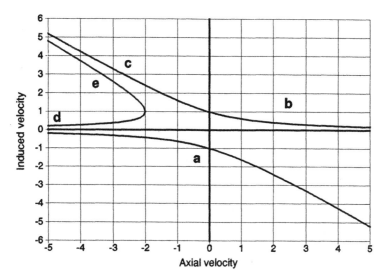

**Fig. 4.3.** Solutions of the induced velocity equations.

portion of the curve lying in the positive climb speed region (normal working state).

(c) This is the part of the curve which is rejected. It defines the part of the flight regime where the notion of a streamtube breaks down (vortex ring state). Because of this, the actuator disc theory cannot be invoked.

(d) This part of the curve relates the flight regime of higher speed descent which allows a streamtube to exist and therefore can be used for the analysis (windmill brake state).

(e) This is the extension of (d) which moves into the vortex ring state where, like (c), it must be rejected (turbulent wake state).

In summary, (b) and (d) only can be used with any degree of confidence. This leaves a gap in the middle of the range where the idea of a streamtube does not conform to the physical behaviour of the rotor wake and experimental data is used to complete the variation of induced velocity with axial velocity. Figure 4.4 shows a typical diagram of this variation with the locations of the rotor states indicated. The flow patterns associated with the rotor states which divide the axial flight regimes from high speed climb to high speed descent are described below.

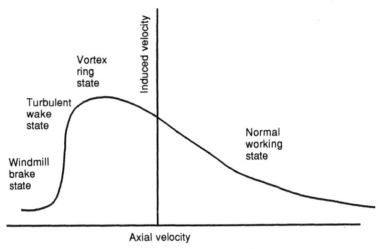

**Fig. 4.4.** Rotor flow states and the average induced velocity.

## Flow patterns surrounding the rotor in axial flight

### Normal working state

This covers climbing and hovering flight. Within the streamtube the airflow relative to the rotor is at all times in a downward direction, and in order to obtain an upward thrust, the air must be accelerated downward giving the wake its characteristic contraction. The notion of a dividing streamtube (separating the flow passing through the rotor disc from the remainder) is realistic and a momentum theory can be sensibly used. The streamtube will in fact consist of vorticity but its concept is pertinent (see Fig. 4.5).

At

$$V_C = 0 \qquad (4.20)$$

the rotor is in hover and the radius of the streamtube above the rotor becomes infinitely large. The reason for this is that the mass of air passing any cross-section of the streamtube must be constant (the air is considered incompressible) and in the hover condition, the velocity falls to zero far upstream. Hover is a limiting case but momentum theory still affords a good estimate of the performance.

### Vortex ring state

As the rotor moves from a hovering condition into a descent, the modelling of momentum theory becomes increasingly difficult to justify. For low descent rates there is a recirculation in the vicinity of the blade tip but there is still downflow in

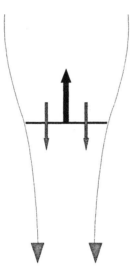

**Fig. 4.5.** Normal working state.

the centre of the disc. The flow widens as it passes the rotor and turbulence appears above the disc. The recirculation and turbulence increase with descent speed and cause the high vibration levels associated with this condition known as the vortex ring state. As the descent speed increases, the recirculation extends from the tip regions to envelop the whole of the rotor disc. The vorticity produced by the rotor blades starts to congregate underneath the rotor disc and periodically detaches from the plane of the disc causing high vibration of low frequency and also control difficulties. The rotor is effectively descending into its own wake and when it approaches a speed equal to the averaged induced velocity over the rotor disc the wake vorticity congregates around the rotor (see Fig. 4.6).

Difficulty in control is exacerbated by the fact that in this flight condition the average induced velocity varies little with descent velocity (see Fig. 4.4), hence the descent rate is hard to maintain. (As the momentum theory breaks down in this rotor state, experimental results must be used to fill in the gap in the theory.) Since the induced velocity varies over the rotor disc, only average values can sensibly be used to assess the flow state within the confines of a momentum theory analysis.

At low descent rates use of the momentum theory can be extended into the vortex ring state, but as the descent rate increases and recirculation moves inward to envelop the entire rotor disc, the momentum theory cannot be justified on physical grounds and the predictions become wholly unsatisfactory.

The limit of the vortex ring state is when the net flow through the rotor disc is zero, i.e.

$$V_D - v_i = 0 \qquad (4.21)$$

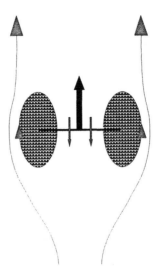

**Fig. 4.6.** Vortex ring state.

## *Turbulent wake state*

This state is so-called because the wake above the rotor is similar to the turbulent wake off a bluff body (see Fig. 4.7). The state is entered when the descent velocity exceeds that required to give the limiting condition of zero net flow through the rotor disc (this condition is sometimes termed "ideal autorotation"). This is equivalent to the rotor being viewed as a flat plate. Since momentum theory only acknowledges induced power, this condition is the power off flight state.

(Although the net flow through the disc is zero, the detailed flow state possesses a reasonable quantity of recirculation and turbulence.)

Flight in this turbulent wake state has a disturbed character due to the turbulent nature of the flow. However, the vortex ring state is usually more severe.

As the descent velocity increases, the net flow at the rotor disc is now upward. The velocity far above the rotor disc given by

$$V_D - 2v_i \qquad (4.22)$$

is still downward but eventually reduces to zero at which point the wake is infinitely wide because of the condition that the mass flow must be constant along the streamtube.

This is the limit of the turbulent wake state. The streamtube concept adopted by the momentum theory model cannot be justified physically for the turbulent wake state and the calculations using a momentum approach are in error.

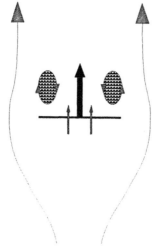

**Fig. 4.7.** Turbulent wake state.

## Windmill brake state

As the descent speed is increased still further, the flow relative to the rotor is upward throughout. A definite streamtube re-establishes itself and momentum theory can be justified physically and applied with confidence (see Fig. 4.8).

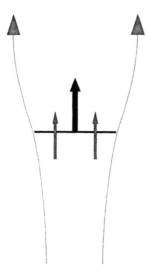

**Fig. 4.8.** Windmill brake state.

## Power required in climb and descent

As the helicopter climbs or descends, the power requirement will alter. The thrust of the rotor will not only have to work against the induced velocity but of that combined with the climb/descent velocity. For instance

$$P_i = T(V_C + v_i) \qquad (4.23)$$

The effect on the rotor performance is shown in Fig. 4.9.

The vertical climb power limit can be easily determined from this figure. A typical value for $V_{C\,max}$ is 5–10 m/s (1000–2000 ft/min) giving $V_C/V_0$ a value of about 1. (To a first order, the profile power does not vary with climb or descent rates.)

An important fact is shown in Fig. 4.9. As the vertical descent velocity increases the power becomes negative. As previously explained the theory breaks down in the vortex ring state so in order to fill in that condition the climb equation (4.17) is used for $-2 < V_C < \infty$, switching to the descent equation (4.18) at the condition

$$V_C = -2 \qquad (4.24)$$

and hence this applies over the range $-\infty < V_C < -2$ as is indicated in the figure. To derive thrust the rotor must produce an induced velocity downwards and therefore incur the penalty of induced power. This is added to the descent velocity which is of negative value, and the combination of the two powers can now become negative. In other words, if the descent velocity exceeds a given value, the power is negative and the rotor is being turned by the airflow. This is the basis of autorotation which is described in Chapter 7 and this technique is used to safely land the

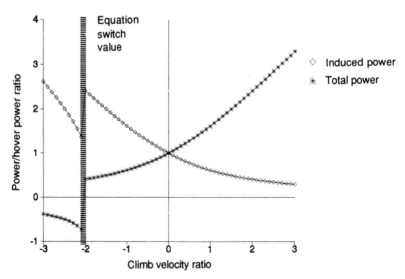

**Fig. 4.9.** Variation of power in vertical climb.

helicopter in the event of partial or complete engine failure. The rotor in axial flight can experience several unwanted problems such as blade dynamic stresses, fluctuating control loads, and rotor speed variation.

## Vertical drag

So far the presence of the fuselage under the rotor disc has been largely ignored. The effect of its presence in axial flight is to interrupt the rotor downwash, creating an unwanted download, and effectively making the rotor produce more thrust than is at first sight required. This extra thrust is termed blockage and to obtain an estimate of its effect in axial flight consider an idealisation of the fuselage. It is modelled by a circular cylinder of diameter $D$, and the region of the downwash affects the fuselage over a length $2R$, the rotor diameter. If we take $C_{D(CC)}$ to be the profile drag coefficient of a circular cylinder the downforce on the fuselage is given by

$$\tfrac{1}{2}\rho(V_C+v_i)^2 \cdot 2R \cdot D \cdot C_{D(CC)} \tag{4.25}$$

using the result from Eqns (4.5) and (4.6)

$$T = 2\rho A(V_C+v_i)v_i \tag{4.26}$$

the ratio of the blockage power to rotor climb power is given by

$$\frac{P_{par}}{P_{climb}} = \frac{\tfrac{1}{2}\rho(V_C+v_i)^3 2R\, D\, C_{D(CC)}}{T(V_C+v_i)} = \frac{1}{\pi}C_{D(CC)}\frac{D}{2R}\left(1+\frac{V_C}{v_i}\right) \tag{4.27}$$

Typical values of the terms are $2R/D$ lies in the range 4–10 and $C_{D(CC)}$ is about 1 giving a power ratio of about 0.05–0.1 in the hover and 0.1–0.2 in maximum climb. Therefore a 5–20% power increase can occur, because of the influence of the fuselage.

If a compound/winged configuration is being analysed, the influence of the wing surfaces in the rotor downwash must be considered. In such circumstances, the value of $C_D$ for the wing structure will be about 2, since it is effectively a flat plate aligned at right angles to the downwash.

The other method of estimating the influence of the fuselage is to apply a factor to the rotor thrust which artificially increases it to give a net thrust equal to that required. This will be used in Chapter 14 where a method for analysing the performance of the complete helicopter is described.

# Ducted rotors

The enclosure of the rotor within a duct has been used in past and existing VTOL aircraft for main lifting rotors (Bell X22A) or tail rotors (SA341 Gazelle). To explain the advantages of ducting the rotors a re-examination of the actuator disc

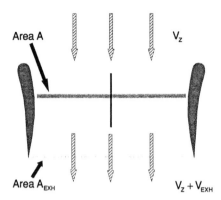

**Fig. 4.10.** Geometry of rotor/duct model.

model is made.[3] The power required of the engines in turning the rotor is equal to the kinetic energy imparted to the flow per unit time. If $\sigma$ is the mass flow through the duct in unit time, $V_Z$ is the axial air velocity into the duct, and $V_{EXH}$ is the axial velocity increase on exit from the duct, then with reference to Fig. 4.10 this gives

$$\text{power} = \tfrac{1}{2}\sigma(V_Z + V_{EXH})^2 - \tfrac{1}{2}\sigma V_Z^2$$

$$= \tfrac{1}{2}\sigma(2V_Z + V_{EXH})V_{EXH} \tag{4.28}$$

From momentum considerations

$$\text{thrust} = \sigma(V_Z + V_{EXH}) - \sigma V_Z$$

$$= \sigma V_{EXH} \tag{4.29}$$

from which

$$\frac{\text{power}}{\text{thrust}} = V_Z + \frac{V_{EXH}}{2} \tag{4.30}$$

Maximum efficiency is therefore gained with a minimum value of $V_{EXH}$, the exhaust velocity. With a free rotor the exhaust velocity is $v_2$, the final wake velocity, and is therefore governed by an area contraction ratio of 2. Hence if the contraction can be reduced, and indeed reversed to an expansion, then the exhaust velocity $V_{EXH}$ is reduced and the power/thrust ratio made more favourable. This is the mechanism which the duct provides and the wake diameter is then constrained to follow the inner contours of the duct.

To complete the analysis we have at the exhaust

$$\sigma = \rho A_{EXH}(V_Z + V_{EXH}) \tag{4.31}$$

from which

$$T = \rho A_{EXH}(V_Z + V_{EXH})V_{EXH} \tag{4.32}$$

where $A_{EXH}$ is the exhaust cross-sectional area.

Defining the area ratio, $\Phi$, to be

$$\Phi = \frac{A}{A_{EXH}}$$ (4.33)

where $A$ is the rotor/fan area, we have for the thrust

$$T = \frac{\rho A}{\Phi}\left(V_z + V_{EXH}\right)V_{EXH}$$ (4.34)

rearranging gives

$$V_{EXH}^2 + V_z V_{EXH} - \frac{T}{\rho A}\Phi = 0$$ (4.35)

from which the power/thrust ratio is

$$\frac{\text{power}}{\text{thrust}} = V_z + \frac{V_{EXH}}{2}$$

$$= V_z + \frac{-V_z + \sqrt{V_z^2 + 4\Phi\dfrac{T}{\rho A}}}{4}$$ (4.36)

$$= \frac{1}{4}\left[3V_z + \sqrt{V_z^2 + 4\Phi\frac{T}{\rho A}}\right]$$

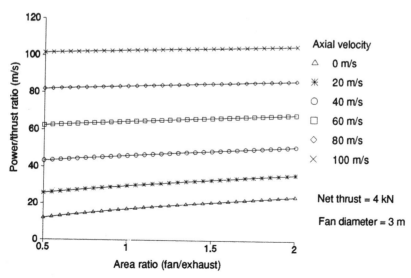

**Fig. 4.11.** Power/thrust ratio variation of ducted rotor.

The variation of power/thrust against $V_z$ and $\Phi$ is shown in Fig. 4.11 and is based on a rotor/duct of diameter 3 m producing 4 kN of thrust. The unducted rotor occurs when $\Phi = 2$. In the hover, the advantage of the duct is apparent and an area ratio duct of 1 has a 41% increase in power/thrust ratio over a free rotor.

With such a saving, the question is posed as to why the ducted rotor has not been used for more than the sporadic nature in the past. To answer this question Fig. 4.11 also shows how the advantage that a lower area ratio gives to the hover condition decreases as the axial velocity of the duct increases. Additionally, if the duct provides an expansion (i.e. $\Phi < 1$) then it is essential that the angle of the duct is not too great as to cause the flow to separate. This will then require a long duct to achieve the expansion required while allowing the flow to remain attached to the duct wall. It is the case, therefore, that the advantage gained at low speed has to pay the price of potentially large ducts with associated weight penalties, whose effectiveness diminishes as the axial speed increases. The advantage gained in hover can therefore inflict weight penalties in cruise conditions.

One final comment is necessary. The analysis of the power was based on the rate of increase of flow kinetic energy with time, rather than the product of rotor thrust and local velocity. This is because the fact that the duct imposes a contraction requires that the duct contribute to the overall performance. In the case of an area ratio ($\Phi$) of 2, the two methods return to the previous analysis of this chapter and the effect of the duct vanishes and the resulting expressions agree.

# 5

# The rotor in forward flight

## Forward flight performance

In hover and vertical flight both the flow through the rotor and the aerodynamic forces on the blades do not vary as the blades rotate around the shaft. However, when the helicopter moves into forward flight, the rotor moves in a basically edge-wise manner and the axial symmetry of flow and blade load variation seen in axial flight is lost. The flow velocities on a general blade section have, as before, the component due to the shaft rotation, but in forward flight there is now an additional component due to the forward flight speed of the complete aircraft. This velocity is in a rearward direction relative to the fuselage which requires resolution into the directions tangential and normal to the blade chordline. An additional consideration is generated since, as the helicopter gains speed, the influence of the fuselage is manifested in the form of an additional drag, known as parasite drag, which must be overcome by the main rotor. In order to overcome the drag of the aircraft, the rotor disc has to be tilted forward by a small amount, which will in turn superimpose an extra velocity component normal to the rotor disc which further complicates the flow patterns through it. The variation of the loads acting on the rotor blades are complex and for the purposes of rotor blade control and vibration assessment must be studied in detail. However, for overall performance assessment, such as thrust and power, average or mean values of the forces are sufficient.

### Induced velocity in forward flight

In evaluating the performance of the rotor in forward flight, the first step is to look at how the principles of a momentum approach must be altered to allow for the forward disc tilt and the two components of velocity. For hover and axial flight the dividing streamtube was obvious in its extent, namely the rotor disc periphery. However, in forward flight this idea has to be modified. For the case of axial flight, the modelling of the streamtube must revert to the disc edge as the forward flight speed reduces to zero. Additionally, the results of the forward flight momentum model must also agree with fixed wing theories as the flight speed increases and in the limit the rotor disc must assume the form of a circular wing. In this latter case the rotational speed can be neglected since the forward flight speed will dominate, and the result to be aimed for is that of a monoplane wing. In the analysis of a monoplane wing (see Glauert[11]) the determining reference area for the downwash velocity is a circle bounded by the entire wing (i.e. whose overall

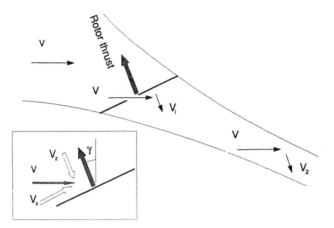

**Fig. 5.1.** Actuator disc in forward flight.

wing span forms a diameter). In order to agree with the limiting cases of axial flight and a monoplane wing, for high speed flight, the control volume for the helicopter rotor must be a sphere with the rotor disc being a generator. With this in mind the actuator disc model of the helicopter rotor in forward flight can now be derived, as shown in Fig. 5.1. In the figure $V$ is the forward speed of the aircraft and $\gamma$ is the forward disc inclination required to overcome the fuselage drag. The rotor thrust is taken as being normal to the disc plane (experience shows that this is justified), which means that the induced velocity is in the reverse direction, i.e. the downward normal to the disc.

The incident flow velocities are resolved in the plane of the rotor disc ($V_x$), and perpendicular to it ($V_z$). The flow velocity in the rotor plane is

$$V_x = V \cos \gamma \tag{5.1}$$

and normal to the disc (including the downwash)

$$V \sin \gamma + v_i = V_z + v_i \tag{5.2}$$

This gives a total flow speed of

$$V' = \sqrt{V_x^2 + (V_z + v_i)^2} \tag{5.3}$$

It is also assumed that far downstream the velocity in the streamtube is composed of the forward speed component to which is added $v_2$ which is equal to twice the induced velocity at the disc (as already seen in the axial flight case). From momentum considerations this gives the rotor thrust as the rate of change of momentum normal to the rotor disc, i.e.

$$T = 2\rho A V' v_i \tag{5.4}$$

The usual way of analysing the solution of the equations is to substitute the total velocity expression (5.3) in the thrust expression (5.4) and isolate $v_i$. This will lead to the following expression where $v_i$ lies on one side of the equation and also within the square root term of the other

$$v_i = \left(\frac{T}{2\rho A}\right)\frac{1}{V'} = \left(\frac{T}{2\rho A}\right)\frac{1}{\sqrt{V_x^2 + (V_z + v_i)^2}} \tag{5.5}$$

Equation (5.5) can be non-dimensionalised by using the tip speed $V_T$ and the thrust coefficient $C_T$. The forward speed $V$ when non-dimensionalised by the tip speed gives a term known as the advance ratio, which is usually denoted by $\mu$. The components $V_x$ and $V_z$ will become $\mu_x$ and $\mu_z$. With the non-dimensional downwash $\lambda_i$ we have the following definitions

$$\mu = \frac{V}{V_T}, \qquad \mu_x = \frac{V_x}{V_T}$$

$$\mu_z = \frac{V_z}{V_T}, \qquad \lambda_i = \frac{v_i}{V_T}$$

$$\mu_{zD} = \frac{(V_z + v_i)}{V_T}$$

$$= \mu_z + \lambda_i \tag{5.6}$$

from which the non-dimensional form of (5.5) becomes

$$\lambda_i = \frac{C_T}{4}\frac{1}{\sqrt{\mu_x^2 + (\mu_z + \lambda_i)^2}} \tag{5.5a}$$

This can be solved by an iterative approach which is discussed below.

The iteration can be a simple, straight substitution as expressed by the equation itself. However, in or near to hover, if a bad choice of starting value is taken, a very slow convergence takes place and in the worst situation, hover, can result in no convergence at all when the iteration will oscillate between two results. For this reason a modification to the expression can be made to collect both terms on one side of the equation and then to invoke the Newton–Raphson technique to determine the solution. This results in a more complicated iterative expression, but the method rapidly converges at all flight conditions and for this reason is generally more robust.

Recasting Eqn (5.5) gives

$$f(v_i) \triangleq v_i - \left(\frac{T}{2\rho A}\right)\frac{1}{\sqrt{V_x^2 + (V_z + v_i)^2}} = 0 \tag{5.7}$$

and applying the Newton–Raphson formula

$$v_i \Leftarrow v_i - \frac{f(v_i)}{\dfrac{\partial f}{\partial v_i}(v_i)} \tag{5.8}$$

gives

$$v_i \Leftarrow v_i - \left[ \frac{v_i - \dfrac{T}{2\rho A}\dfrac{1}{\sqrt{V_x^2 + (V_z + v_i)^2}}}{1 + \dfrac{T}{2\rho A}\dfrac{V_z + v_i}{\sqrt{V_x^2 + (V_z + v_i)^2}^3}} \right] \tag{5.9}$$

and in non-dimensional form

$$\lambda_i \Leftarrow \lambda_i - \left[ \frac{\lambda_i - \dfrac{C_T}{4}\dfrac{1}{\sqrt{\mu_x^2 + (\mu_z + \lambda_i)^2}}}{1 + \dfrac{C_T}{4}\dfrac{\mu_z + \lambda_i}{\sqrt{\mu_x^2 + (\mu_z + \lambda_i)^2}^3}} \right] \tag{5.9a}$$

As an alternative, Eqn (5.5) could be multiplied out to form a quartic in $v_i$, but this is rather unwieldy and because of its ease of implementation, the iterative technique is usually applied.

To show the effect of forward speed on induced velocity, Fig. 5.2 shows the various power components for the main rotor of a typical helicopter. The induced velocity will control the induced power, and from Fig. 5.2 the most immediate and

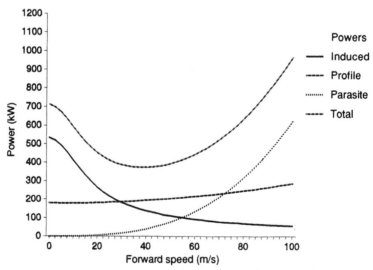

**Fig. 5.2.** Power variation in forward flight. Main rotor performance.

important feature shown is the rapid fall-off of the induced power with forward speed. The induced velocity, therefore, reduces significantly in magnitude as the forward flight speed of the helicopter increases. This means that the induced power, which is the major factor in the hover, very quickly loses its stranglehold on the rotor performance as the aircraft accelerates away from the hover. The rotor will indeed require extra power to overcome the increased effect of the profile drag of the blades but will now also require an extra power component, since the rotor has to drag the fuselage through the air. In hover, the fuselage has an influence in that it interrupts the downflow from the rotor, however in forward flight the influence of the fuselage is much greater, especially at high speed and, as can be seen in the figure, ultimately becomes the dominant factor as regards power consumption. This is particularly true of the types of fuselage shapes used for helicopters.

In forward flight the power required to overcome the fuselage drag (called the parasite power) can be obtained by examining the rate of work done by the rotor. It is producing a thrust $T$ with a total flow speed, through the rotor disc, in the negative thrust direction of $(V_z + v_i)$ and thereby requiring a power of

$$P = T(V_z + v_i) \tag{5.10}$$

The first term is the parasite power and it can be seen that the second refers to the induced component. Figure 5.2 shows a steady increase in the profile power which will be discussed in the next section.

## Blade element considerations

In previous chapters the performance of the rotor in hover was evaluated by use of blade element theory. This technique will now be extended to forward flight. The construction of the equations follows the previous analysis, but in forward flight there are now two sources of air velocity over a blade section, namely, that due to the rotation of the blades about the shaft and that due to the forward speed of the entire helicopter. If we only consider the velocity normal to a blade section then the first velocity component, $\Omega r$, is already normal to the blade. However the forward flight speed has to be resolved into the direction normal to the blade (see Fig. 5.3).

Firstly, the forward velocity is resolved into components parallel and normal to the disc, $V_x$ and $V_z$, respectively. The component of $V_x$ resolved normal to the blade span will vary according to the position of the rotor blade relative to the flight direction. The azimuthal position is conventionally expressed as the angle subtended between the blade and the datum which is usually along the tail boom. The angle is positive in the direction of blade rotation. The forward speed component normal to the blade is then given by the component of forward velocity parallel to the rotor disc, $V_x$, resolved normal to the blade span, namely

$$V_x \sin \psi \tag{5.11}$$

from which the total velocity component normal to the blade span is

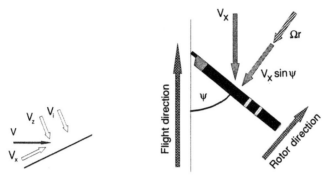

**Fig. 5.3.** Velocity components in forward flight.

$$\Omega r + V_x \sin \psi \qquad (5.12)$$

this is denoted by $U_T$.

It is again usual to non-dimensionalise the velocities by the rotor tip speed, and this reduces (5.12) to

$$U_T = V_T \left( x + \frac{V_x}{V_T} \sin \psi \right) \qquad (5.13)$$

and recalling

$$\lambda_i = \frac{v_i}{V_T} \qquad (5.14)$$

we find the velocity components shown in Fig. 5.4 are given by

$$U_T = V_T (x + \mu_x \sin \psi)$$
$$U_P = V_T (\mu_z + \lambda_i) \qquad (5.15)$$

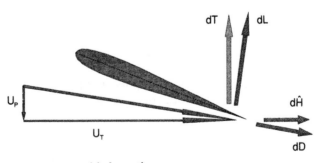

**Fig. 5.4.** Force components on blade section.

This gives the total flow velocity over the blade section as

$$V = \sqrt{U_P^2 + U_T^2}$$
$$= V_T \sqrt{(x + \mu_x \sin \psi)^2 + (\mu_z + \lambda_i)^2} \qquad (5.16)$$

Due to orders of magnitude, the term $\mu_z + \lambda_i$ may sensibly be neglected. Hence

$$V \approx V_T(x + \mu_x \sin \psi) \qquad (5.17)$$

giving the inflow angle as

$$\phi = \tan^{-1} \left( \frac{V_T(\mu_z + \lambda_i)}{V_T(x + \mu_x \sin \psi)} \right) \qquad (5.18)$$

for small angles we can approximate this to

$$\phi = \frac{\mu_z + \lambda_i}{x + \mu_x \sin \psi} \qquad (5.19)$$

the angle of incidence now becomes

$$\alpha = \theta - \frac{\mu_z + \lambda_i}{x + \mu_x \sin \psi} = \theta - \frac{\mu_{zD}}{x + \mu_x \sin \psi} \qquad (5.20)$$

The forces on the blade section are then

$$\text{(lift)} \quad dL = \tfrac{1}{2} \rho V^2 c \, dr C_L$$
$$= \tfrac{1}{2} \rho V^2 c \, dr a \alpha \qquad (5.21)$$
$$\text{(drag)} \quad dD = \tfrac{1}{2} \rho V^2 c \, dr C_D \qquad (5.22)$$

The two forces, $dL$ and $dD$, are then resolved into the rotor thrust direction $dT$, and that direction in the rotor disc plane and normal to the blade span $d\hat{H}$

$$\text{(thrust)} \quad dT = dL \cos \phi - dD \sin \phi \approx dL \qquad (5.23)$$

This simplification is justified since with an unstalled aerofoil section the lift term $dL$ is substantially greater than the drag $dD$. Also, as $\phi$ is small, $\cos \phi \approx 1$ and $\sin \phi \approx \phi$, therefore the first term in (5.23) is greater than the second by two orders of magnitude and can be neglected

$$\text{(in-plane force } (\hat{H})) \quad d\hat{H} = dL \sin\phi + dD \cos\phi \simeq dL\phi + dD \quad (5.24)$$

In this case, both terms are of similar magnitude and must be retained

$$\text{(torque)} \quad dQ = d\hat{H}\, r \quad (5.25)$$

The blade pitch angle, $\theta$, has now to be specified. It is a combination of collective and cyclic pitches and is expressed in terms of the azimuth angle $\psi$. Collective pitch is constant with respect to azimuth and cyclic pitch varies at the rotor frequency (i.e. once per revolution) which means that the blade pitch can be expressed as the first three terms of a Fourier series and due to historical reasons, it is a negative Fourier series

$$\theta = \theta_0 - A_1 \cos\psi - B_1 \sin\psi \quad (5.26)$$

$\theta_0$ is the collective pitch and has already been discussed. $A_1$ is the cyclic pitch component which is applied longitudinally with respect to the fuselage, and $B_1$ the corresponding laterally applied component (see Plate 5.1).

If we assume constant chord, zero blade twist and the blade pitch angle expressed by the collective and $B_1$ cyclic terms only, i.e.

$$\theta = \theta_0 - B_1 \sin\psi \quad (5.27)$$

We can now evaluate the performance of the complete rotor.

As will be shown in Chapter 6, $B_1$ cyclic pitch is applied laterally but gives a longitudinal rotor disc tilt, and therefore $B_1$ is normally referred to as the longitudinal cyclic pitch. Similarly $A_1$ is called the lateral cyclic pitch because it gives a lateral disc tilt even though it is applied longitudinally. See equations (6.34) and (6.35).

**Plate 5.1.** EH101 prototype PP5 in forward flight. Note the cyclic pitch giving a higher blade pitch angle on the retreating (far) side than on the advancing (near) side.

In forward flight, both the velocity terms and the pitch terms vary around the disc with the forces varying likewise. In the case of rotor thrust, torque and drag the steady value of those forces are required and so to calculate the performance, the expressions for the forces on the blade sections are integrated along the blade span, as before, but then are averaged around the rotor azimuth. This is achieved by integrating the expressions around the azimuth and dividing the resulting integral by $2\pi$. To avoid confusion, those quantities which are averaged will be denoted by an overbar.

The blade pitch will also vary along the blade span if twist is present.

Mean thrust $\overline{T}$

$$\overline{T} = \frac{N}{2\pi}\int_0^{2\pi}\mathrm{d}\psi\int_0^R\frac{\mathrm{d}T}{\mathrm{d}r}\,\mathrm{d}r \tag{5.28}$$

$$C_T = \frac{1}{\frac{1}{2}\rho V_T^2\pi R^2}\frac{N}{2\pi}\int_0^{2\pi}\mathrm{d}\psi\int_0^1\frac{1}{2}\rho V_T^2 caR(x+\mu_x\sin\psi)^2\alpha\,\mathrm{d}x \tag{5.29}$$

$$C_T = \frac{sa}{2}\left[\theta_0\left(\frac{2}{3}+\mu_x^2\right)-\mu_xB_1-\mu_{zD}\right] \tag{5.30}$$

Mean rolling moment $\overline{M}_R$

$$\overline{M}_R = \frac{N}{2\pi}\int_0^{2\pi}\mathrm{d}\psi\int_0^R\frac{\mathrm{d}T}{\mathrm{d}r}r\sin\psi\,\mathrm{d}r \tag{5.31}$$

$$C_{MR} = \frac{\overline{M}_R}{\frac{1}{2}\rho V_T^2\pi R^2 R}$$

$$= \frac{sa}{2}\left[\frac{2}{3}\theta_0\mu_x-\frac{B_1}{4}\left(1+\frac{3}{2}\mu_x^2\right)-\frac{1}{2}\mu_x\mu_{zD}\right] \tag{5.32}$$

Mean torque $\overline{Q}$

$$\overline{Q} = \frac{N}{2\pi}\int_0^{2\pi}\mathrm{d}\psi\int_0^R\frac{\mathrm{d}\hat{H}}{\mathrm{d}r}r\,\mathrm{d}r \tag{5.33}$$

$$C_Q = \frac{\overline{Q}}{\frac{1}{2}\rho V_T^2\pi R^2 R}$$

$$= s\left[\mu_{zD}a\left(\frac{\theta_0}{3}-\frac{1}{2}\mu_{zD}\right)-\frac{1}{4}\mu_x\mu_{zD}aB_1+\frac{C_D}{4}\left(1+\mu_x^2\right)\right] \tag{5.34}$$

which on rearranging becomes

$$C_Q = \frac{sa}{2}\mu_{zD}\left[\frac{2}{3}\theta_0 - \mu_{zD} - \frac{1}{2}\mu_x B_1\right] + \frac{s}{4}C_D\left[1+\mu_x^2\right] \tag{5.35}$$

Mean rotor drag (H force)

$$H = \frac{N}{2\pi}\int_0^{2\pi} d\psi \int_0^R \frac{d\hat{H}}{dr}\sin\psi \, dr \tag{5.36}$$

$$C_H = s\left[a\mu_{zD}\left(\frac{\mu_x}{2}\theta_0 - \frac{1}{4}B_1\right) + \frac{1}{2}\mu_x C_D\right] \tag{5.37}$$

The rotor must overcome two sources of power drain, namely the mean torque from driving the rotor (profile), obtaining thrust (induced), and the need to over-come the extra drag due to the rotor (H force). This drag is in addition to the parasite drag of the fuselage. The power required for this is given by

$$C_{Q+H} = \frac{\overline{Q}\Omega + V_T\mu_x H}{\frac{1}{2}\rho V_T^2\pi R^2\Omega R} \tag{5.38}$$

$$= C_Q + \mu_x C_H$$

$$C_{Q+H} = \frac{sa}{2}\mu_{zD}\left(\frac{2}{3}\theta_0 - \mu_{zD} - \frac{1}{2}\mu_x B_1\right) + \frac{s}{4}C_D\left(1+\mu_x^2\right)$$

$$+ \mu_x s\left[a\mu_{zD}\left(\frac{1}{2}\mu_x\theta_0 - \frac{1}{4}B_1\right) + \frac{1}{2}\mu_x C_D\right] \tag{5.39}$$

which on rearranging becomes

$$C_{Q+H} = \mu_{zD}\frac{sa}{2}\left[\theta_0\left(\frac{2}{3}+\mu_x^2\right) - \mu_x B_1 - \mu_{zD}\right] + \frac{C_D}{4}s\left(1+3\mu_x^2\right) \tag{5.40}$$

$$= \mu_{zD}C_T + \tfrac{1}{4}C_D s\left(1+3\mu_x^2\right)$$

The result is the sum of two terms. The first represents the product of the rotor thrust and the total inflow velocity normal to the rotor disc and matches the result previously derived. The second term represents the profile power consumption. Examination of this term and its derivation shows that the formula for the profile power in hover is repeated but with a factor of $1+3\mu_x^2$. This consists of $1+\mu_x^2$ for the rotational velocity effects and $2\mu_x^2$ for the rotor drag (H force).

The following shortcomings of the analysis must be noted:

1.  No allowance has been made of the advancing blade problem of high Mach number leading to drag rise and other compressibility problems.
2.  No consideration has been given to the high pitch angles which will be borne by the retreating blade at high advance ratio, giving rise to stall problems.
3.  The circular reverse flow region on the retreating side of the disc, defined by $U_T \leqslant 0$, has been ignored. This will cause performance losses as the advance ratio increases, commensurate with the helicopter flight speed. Inspection of (5.15) shows that this region has its centre at $(x = \frac{1}{2}\mu_x$ and $\psi = 270°)$ and has radius of $\frac{1}{2}\mu_x$.
4.  The velocity perturbations caused by rotor blade motion have been ignored.
5.  The inflow normal to the rotor disc has been assumed to be uniform. In reality it is very complicated, especially at high speed. This fact is the source of much research and investigative work which has already produced many calculation procedures. A simple model of the rotor induced flow in forward flight was proposed, many years ago, by H. Glauert of RAE Farnborough. He visualised the rotor wake as a vortex tube skewed to represent the combination of the downflow through the disc and the forward speed component in the plane of the disc. This produces an inflow distribution which is close to a plane inclined forwards but symmetric laterally. For a description of this type of wake see Castles and De Leeuw.[12] The inflow for such a distribution can be expressed, with an approximation, by the formula

$$v_i = v_{i0}\left(1 + E\frac{r}{R}\cos\psi\right) \tag{5.41}$$

The forward inclination of the distribution is determined by the value of $E$, which is known as the Glauert factor. Its simplicity means that it can be used in an aerodynamic analysis of a helicopter rotor as the next step on from a uniform downwash distribution. If a detailed aerodynamic model is used, from the outset, in a rotor performance calculation, the computer time can be excessive. Use of a Glauert downwash will aid the efficient use of a computer model allowing it to trim a rotor closer to its true situation before a more difficult and time-consuming vortex wake model is invoked. It must be said that in earlier times when computing was relatively pedestrian such comments were undoubtedly true. Modern computer hardware and software have progressed so far that such niceties may not now be so important.

6.  The flow over a blade section has a spanwise component as well as the chordwise component to which we have confined the analysis. Stepniewski[13] investigates the effect on profile power of the effect due to the spanwise flow and, in addition, the reverse flow region.

The results are that the effective profile drag coefficient for the $H$ force component (the $2\mu_x^2$ term) varies in a manner as shown in the reference which requires a factor to be applied. A typical high-speed case where $\mu = 0.35$ gives a 1.5 factor to this term thus making it $3\mu_x^2$. A similar analysis applied to the rotational component of profile power (the $1+\mu_x^2$ term) produces an additional increase of $0.7\mu_x^2$ giving $1+1.7\mu_x^2$. If we combine the above two factors, we obtain the modified expression $1+4.7\mu_x^2$. The 4.7 value is called the Stepniewski factor.

## Stall limitation on thrust (rigid rotor)

The previous analysis shows the variation of roll moment in forward flight with the application of collective and longitudinal cyclic pitch ($\theta_0$ and $B_1$, respectively). In order for the helicopter to fly in trim, the roll moment must be equal to zero, i.e.

$$C_{MR} = 0 \qquad (5.42)$$

Referring to the roll moment expression (5.32), this results in

$$\frac{2}{3}\theta_0\mu_x - \frac{B_1}{4}\left(1+\frac{3}{2}\mu_x^2\right) - \frac{1}{2}\mu_x\mu_{zD} = 0 \qquad (5.43)$$

Ignoring second order terms in $\mu$, this simplifies to

$$B_1 \simeq \frac{8}{3}\mu_x\theta_0 \qquad (5.44)$$

This gives the amount of cyclic pitch required to trim the helicopter in roll, which as previously mentioned will discard lift on the advancing side and increase lift on the retreating side. For a given collective pitch setting, as the forward speed increases the cyclic pitch required to trim the helicopter also increases with the effect of increasing the aerodynamic loading on the retreating side of the disc. At the azimuth position $\psi = 270°$, the retreating blade pitch angle is given by $\theta_0 + B_1$ which will have an upper bound value equal to the stalling angle of the aerofoil section. If we denote this by $\theta_s$, we must have, for the limiting case

$$\theta_0 + B_1 = \theta_s \qquad (5.45)$$

So in order to derive the maximum thrust from the rotor, we must limit the retreating blade pitch to avoid stall. Consequently, as the cyclic pitch required to trim the helicopter increases, the maximum collective pitch will reduce accordingly. This will have the effect of reducing the rotor thrust potential with increasing speed, and we obtain the classic retreating blade stall limit.

If the expressions for cyclic pitch to trim and the retreating blade stall limit are combined, the following expressions for collective and longitudinal cyclic pitch result

$$\theta_0 = \frac{\theta_s}{1+\frac{8}{3}\mu_x}, \qquad B_1 = \frac{8}{3}\mu_x\frac{\theta_s}{1+\frac{8}{3}\mu_x} \qquad (5.46)$$

Substitution of these results in the rotor thrust expression (5.30) gives rise to the following

$$C_T = \frac{sa}{2}\left[\left(\frac{2}{3}+\mu_x^2\right)\frac{\theta_s}{1+\frac{8}{3}\mu_x} - \mu_x\frac{8}{3}\mu_x\frac{\theta_s}{1+\frac{8}{3}\mu_x}\right]$$  (5.47)

i.e. to first order $\mu$

$$C_T \simeq \frac{sa}{3}\frac{\theta_s}{1+\frac{8}{3}\mu_x}$$  (5.48)

[There is a term $(\mu_{zD})$ in the thrust expression which has been ignored for reasons of clarity. This analysis is to show the thrust limit behaviour in its simplest form, and ignoring this term does not invalidate the conclusions.] This analysis shows that the requirement for roll trim, together with the need to avoid stall on the retreating blade, causes the rotor thrust capability to decrease with increasing forward flight speed. This effect is plotted in Fig. 5.5.

The previous analysis does not consider the advancing blade whatsoever. To correct this, a similar analysis can be performed on the advancing blade tip where a stalling pitch angle at the advancing tip is specified. The resulting expression for rotor thrust is then given by

$$C_T \simeq \frac{sa}{3}\frac{\theta_s}{1-\frac{8}{3}\mu_x}$$  (5.49)

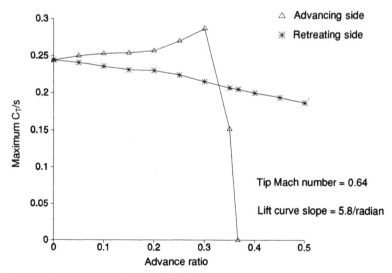

**Fig. 5.5.** Rotor thrust limits in forward flight.

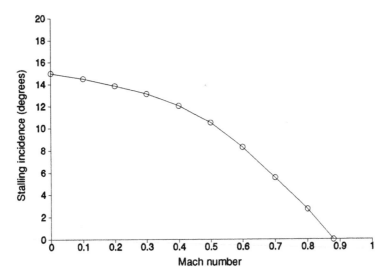

**Fig. 5.6.** Aerofoil section boundary.

On the face of it the sky is the limit as the value of the expression increases with forward speed. It should be noted that this expression tends to infinity as $\mu_x = 0.375$ is approached which is an anomaly resulting from simplifications of the analysis. In fact, as the forward speed increases the advancing blade tip begins to experience compressibility effects and in order to avoid drag rise or shock induced stall the advancing blade tip must be restricted in pitch. The limitation on pitch for a representative aerofoil section varies with Mach number, an example of which is

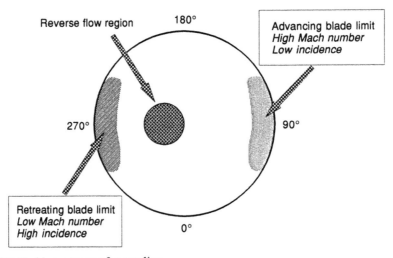

**Fig. 5.7.** Problems areas of rotor disc.

**Table 5.1**

| Helicopter component | Drag (% of total) |
|---|---|
| Fuselage | 13.0 |
| Pylon | 3.0 |
| Hub† | 20.0 |
| Skids | 8.0 |
| Antennae, etc. | 13.0 |
| Induced drag of above (3° nose down) | 2.5 |
| Sideslip of above (3°) | 5.0 |
| Main rotor controls | 3.0 |
| Tail rotor (hub and controls) | 12.5 |
| Leakage, engines, etc. | 12.5 |
| Transmission cooling | 5.0 |

†Of particular note is the contribution of the hub, which is attracting much attention in the aerodynamic streamlining of helicopters, an example being the gloves fitted to the main rotor of the Westland Lynx (G-LYNX) for its world speed record attempt, and the EH101 *Merlin* prototype PP5.

shown in Fig. 5.6. If this limitation is applied to the advancing blade thrust limit, via the $\theta_s$ term, the variation of maximum thrust is also shown in Fig. 5.5.

Figure 5.5 shows the combined limits to the rotor thrust from each side of the disc. The main limit comes from the retreating blade, but at high speed, the advancing blade limit causes an abrupt cut-off. Taking both limits together, the rotor is boxed in by the two sets of lines.

There are then three problem areas of a rotor disc in forward flight namely, the advancing and retreating tip regions and the reverse flow region as shown in Fig. 5.7. The retreating blade limit has been a constant problem of the edgewise rotor with various methods of solution emerging. One of these has already been described and is the Advancing Blade Concept (ABC™)[2] used in the Sikorsky S69 aircraft. Here the need for roll trim is removed by using a coaxial rotor arrangement, and achieving the required trim by symmetry alone, whereby each individual rotor is not trimmed in roll but the roll moment of each is cancelled out by the symmetry of the configuration. The retreating blade limit is now removed and the rotor thrust can now be increased. However, the advancing blade limit still applies.

# Parasite power and drag

So far in the analysis we have examined the induced and profile powers in detail but not the parasite power. Parasite power is incurred because of the need of the main rotor to overcome the drag of the fuselage itself. The drag of this type of body increases with the square of the forward speed, and combining this with the speed, the power possesses a cubic variation. In consequence, parasite power is small at low speed where the induced power is dominant, but at the higher speeds can increase dramatically and become the dominant factor. A typical variation of power component with forward speed is shown in Fig. 5.2.

The drag of the fuselage has many components, examples of which are listed in Table 5.1 together with an idea of their contribution to the total drag force, expressed as a percentage.

# 6

# Dynamics of the rotor

## Introduction

If a helicopter rotor has blades rigidly attached to the hub then in edgewise flight (i.e. when the rotor plane is virtually parallel to the flight direction) it experiences a tendency to roll onto its back with increasing forward speed. This is due to the dissymmetry of lift on either side of the disc (advancing and retreating) creating a turning moment about the hub. In order to avoid this damaging effect the rotor blades must be prevented from transferring the lift moment on the blades to the rotor hub and consequently to the complete aircraft itself. This is accomplished by allowing the rotor blades to "flap". Flapping is a blade motion out of the plane of rotation, and is of paramount importance to the control of a helicopter rotor. It is achieved by attaching the blades to the hub via hinges or flexible elements whose axes lie in the hub plane and allow the blades to rotate in a direction perpendicular to it. The incorporation of flapping hinges in the rotor hub isolates the rolling moment to the blades themselves and so avoids the rolling tendency of the complete airframe in forward flight. However, the rotor blades are now free to move under the joint influences of the aerodynamic forcing and the centrifugal forces acting on the blades in a radially outward direction. The magnitude of the radial forces can be appreciated when we observe that a helicopter main rotor blade tip experiences a radial acceleration of the order 500*g*, whilst a tail rotor blade tip will experience approximately six times that amount. In addition to avoiding the rolling moment, the inclusion of blade flapping hinges or flexures, helps to reduce the vibration felt in the fuselage, cockpit or control system since the path for transmitting the forces and moments from the rotor has been interrupted. Without any further provision of control the blades are free to move according to the relative magnitudes of the aerodynamic and centrifugal forces.

As the helicopter moves into forward flight the advancing blade will experience an increase in the aerodynamic forces (due to an increased dynamic head) and will therefore tend to accelerate upwards out of the disc plane. The converse will happen with the retreating blade which will tend to accelerate downwards because of the reduced dynamic head. Although the blade's maximum upward or downward acceleration occurs on the lateral parts of the disc (i.e. 90° or 270° azimuth), where the dynamic head is at its extreme values, the blades will require a time period to climb or descend to their maximum flapwise deflections. As will be seen later in this chapter this time delay corresponds to an azimuth travel of close to 90°. (A teetering rotor will have exactly 90°.) This will cause the blade to flap up at the front of the disc and down at the rear, i.e. a backward disc tilt will occur. Since the

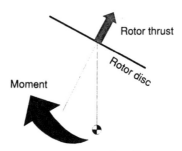

**Fig. 6.1.** Creation of fuselage moment by rotor disc tilt.

thrust direction is effectively normal to the disc plane it will therefore have a rear-ward component and the aircraft will start to decelerate.

The backward rotor disc tilt also causes the line of action of the thrust to move forward of the aircraft CG, creating a nose-up pitching moment which will cause a corresponding change in the helicopter's pitch attitude. This effect is shown in Fig. 6.1. This means that a freely flapping rotor will move in a way to oppose the intended direction of travel. In order to fly in a sensible manner the disc must be tilted forward in forward flight to enable a forward component of thrust to be generated which will maintain the aircraft forward motion by balancing the drag of the entire aircraft. As the natural tendency of the rotor disc is in opposition to this, use of the aerodynamic lift on the blades must be adjusted to force the blades to move in the opposite manner to that induced by the tangential variation of dynamic head. The effect required is an increase in blade pitch on the retreating side driving the blade to flap up at the rear of the disc and a decrease in the lift on the advancing side causing the blades to flap down at the front of the disc. In other words we need to drive the aerodynamics in a periodic (or cyclic) manner which can only be achieved by altering the lift coefficient of the blades as the dynamic head is already determined by the forward flight and rotor speeds. For normal blades this will require use of blade pitch, whilst in later designs of more futuristic types by means of circulation control. We will only consider the former where use of *cyclic pitch* is used as the control mechanism for disc attitude.

The ability to control the attitude of the rotor disc allows the pilot to orientate the main rotor thrust vector and in consequence the attitude of the fuselage. The effectiveness of this mechanism is dependent on the attachment of the rotor blades to the hub and will be discussed in due course. Controlling the attitude of the fuselage requires that a moment be applied about the CG and the simplest way is to move the line of action of the thrust force away from the CG and hence create the moment as already shown in Fig. 6.1. It can be seen that this moment is the product of the rotor thrust and the disc tilt. So to create a given moment a smaller rotor thrust will require a larger disc tilt.

The pilot has two controls for the main rotor, namely the collective pitch where the pitch of the blade is constant around the azimuth and provides the thrust control for the rotor; and the cyclic pitch where the blade pitch is cyclically varied as it rotates around the azimuth in a once per revolution manner causing the rotor disc to tilt, whilst keeping the rotor shaft fixed. The inclusion of flapping motion also carries a penalty. The phenomenon of Coriolis acceleration causes a forcing to be generated in a tangential direction when a rotating system experiences a radial motion. It is like skaters pulling in their arms during a spin. The inward motion of the skaters' arms forms the radial motion which is in a rotating frame due to the spinning. A force is generated in a tangential direction which gives an accelerating torque to the skaters and an increase in the rate of spin is achieved. As a rotor blade flaps up or down part of the motion will be in a radial direction. The Coriolis forcing will then drive the blades in an in-plane manner. If no allowance for this is made, the loading in the blade will cause damage to the root of the blade or the hub and so in order to avoid this, the blades are allowed to move in the disc plane by incorporating hinges or flexures. This motion is termed leading or lagging and can cause additional problems of itself like ground resonance which will be discussed in Chapter 12.

## Application of blade control

The blade rotates in pitch about a bearing, aligned in a radial direction, which can be a roller bearing stack or a composite flexure. The pitch is applied via an arm projecting forwards from the pitch bearing housing known as the pitch horn. (The pitch bearing housing part of the blade is often referred to as the blade cuff.) The pitch horn is connected to its own individual track rod by a swivel bearing and vertical movement of the track rod will cause the change in blade pitch angle. The lower end of the track rod is connected to a spider or rotating star which is constrained to rotate with the rotor. This layout is shown in Fig. 6.2. A movement of the spider in a direction parallel to the rotor shaft will cause all of the blades to rotate in pitch by the same amount, have the same pitch angle change, and hence have a change in collective pitch. If the spider centre maintains its location relative to the rotor shaft but its plane tilts, then it can be seen that as the blade rotates around the shaft, as the rotor turns, the spider arm moves up and down once per rotor revolution. That is, the blade pitch angle changes once per revolution and

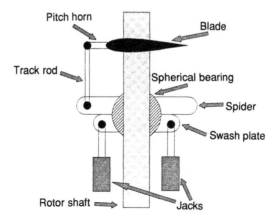

**Fig. 6.2.** Main rotor blade control system—swash plate/spider.

cyclic pitch is achieved. The pilot's controls must therefore be able to control the spider's location and orientation with the added complication of the spider is itself rotating with the shaft. The majority of helicopters achieve this using a swash plate or stationary star. This is essentially a flat plate joined to the spider such that they remain locked together in the same plane. The swash plate and spider combination slides up and down the rotor shaft and tilts relative to its common centre. The swash plate is held stationary relative to the fuselage and its position and orientation is determined by three actuators, or jacks, connecting it to the top of the fuselage or main rotor gearbox casing. The pilot's controls alter the strokes of the actuators which position the swash plate. The spider is constrained to lie in the same plane as the swash plate but rotates with the rotor and its position determines the collective and cyclic pitch angles. If the actuators move in unison, the swash plate and spider maintain any tilt but slide along the rotor shaft and collective pitch is adjusted. If the actuators move unequally then the rotation plane of the swash plate and spider combination is altered and cyclic pitch is achieved. This type of mechanism is shown in Figs 6.2 and 6.3.

Some helicopters use what is termed a dangleberry system for rotor control where the spider location is determined by a shaft which lies within the rotor shaft which is necessarily hollow (see Plate 6.1). The spider rotates about the controlling shaft by means of a bearing and lies within the rotor hub where each spider arm projects though a slot in the rotor hub central casing. The movement of the spider arms requires the slot to be considerably larger than the arms and to seal the mechanism from outside influences, a flexible gaiter is normally fitted. Collective pitch is obtained by a movement of the controlling shaft parallel to the rotor shaft axis and a sideways movement of its lower end will alter the rotation plane of the spider allowing cyclic pitch to be generated. This type of pitch control is shown in Fig. 6.4.

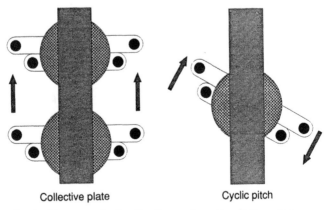

Collective plate                    Cyclic pitch

**Fig. 6.3.** Swash plate/spider control of collective and cyclic pitch angles.

**Plate 6.1.** Westland WG30 semi-rigid main rotor head showing the flapping flexures, pitch horns, dangleberry control system and head vibration absorber. The large amount of wiring is because the aircraft is set up for testing. After development this would not normally be present. (Courtesy Westland Helicopters.)

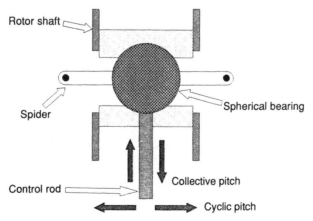

**Fig. 6.4.** Blade pitch control by the dangleberry system.

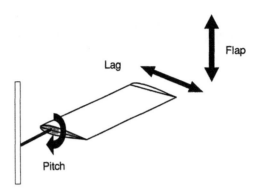

**Fig. 6.5.** Hinge articulation.

# Types of rotors

When the flap and lag freedoms are provided for by inclusion of hinges the rotor is said to be articulated. The hinges are flap, lag and pitch (collective and cyclic). Figure 6.5 shows the blade degrees of freedom described. The order of their installation on the rotor hub can vary from helicopter to helicopter, and this order will have important effects on the dynamic behaviour of the blades themselves.

### Teetering or see-saw rotors

These are the simplest form (Fig. 6.6) and consist of a pair of blades joined at the hub. The hub is free to move in a flapping sense with the flapping hinge line lying on the rotor shaft axis. A good example of this type of rotor hub is found on the Bell 47J aircraft. Each blade has its own pitch hinge (see Plate 6.2).

Some designs of a teetering rotor help minimise the Coriolis forcing by *underslinging* the rotor hub (see Fig. 6.7). With this, the hub flap pivot is positioned above the line of the blades. The reason for its use is as follows. The rotor blades have a built-in steady flap angle (precone) to relieve the roots of bending stress. Underslinging moves the blade/hub CG closer to the flap pivot so as the

**Fig. 6.6.** Flap–pitch, Bell Jet Ranger.

**Plate 6.2.** The teetering rotor of the Bell 47J showing the "bell bar" system. As shown in the photograph, the "bell-bar" is placed underneath the main rotor and perpendicular to the blades. It is a rigid rod rotating with the rotor and able to pivot about its own teetering hinge. It has bob weights fixed to the extremities. The control system passes blade pitch changes to the rotor via this bar which, because of its bob weights, acts as a stabiliser. The linkage from the bar to the pitch horns keeps the rotor plane parallel to the rotation plane of the bar. If the aircraft is disturbed and the rotor shaft tilts, the gyroscopic inertia of the bar maintains its plane of rotation thereby imparting stability to the rotor. (Courtesy Westland Helicopters.)

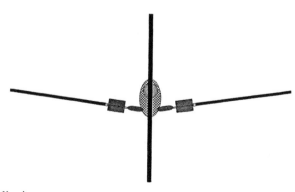

**Fig. 6.7.** Underslinging.

**Plate 6.3.** The articulated main rotor head of a Sikorsky S-65A showing the pitch change mechanism, i.e. pitch horn–track rod–spider–swash plate.

hub and blades rotate about the flapping pivot there is a minimal change in the CG position away from the rotor shaft axis. This modification has a drawback since underslinging brings the problem of *mast bumping*. At low rotor thrust, when a large disc tilt is required for manoeuvre, the motion of the hub can cause an impact of the hub with the rotor mast. This can be accompanied by a fracture of the mast and the loss of the helicopter. Progressive high rate springs can be fitted to the rotor hub to help the aircraft avoid this problem.

## Articulated rotors (Plate 6.3)

For more than two blades the rotor hub must provide each blade with its own flap, lag and pitch hinges. The order of these varies with the particular helicopter. Examples of such variations are shown below. Hinge order is usually expressed in the order of location, moving from the rotor shaft to the blade root.

*Westland Scout* and *Aerospatiale Alouette*. Flap–lag–pitch (Fig. 6.8; note that pitch is often referred to as torsion).

*Boeing Vertol Chinook* and *Aerospatiale Gazelle*. Flap–pitch–lag (Fig. 6.9).

*Sikorsky S58* and *Westland Sea King*. Lag–flap–pitch (Fig. 6.10; the lag and flap hinges are coincident in location on the rotor hub).

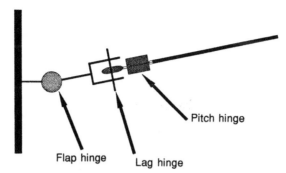

**Fig. 6.8.** Flap–lag–pitch, Westland Scout, Aerospatiale Alouette.

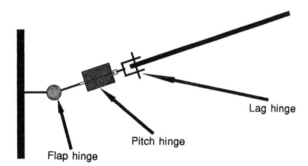

**Fig. 6.9.** Flap–pitch–lag, Boeing Vertol Chinook, Aerospatiale Gazelle.

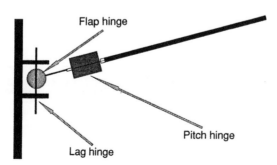

**Fig. 6.10.** Lag–flap–pitch, Westland Wessex, Westland Sea King.

## *Pitch–flap coupling*

It is usual with main rotors to configure the geometry of the hinges in such a way as to remove as much as is possible any kinematic couplings between flap, lag or pitch motions. A good example of this is the pitch angle control horn on a blade which transmits the pitch change to the blade from the track rods connected to the control system below. It is the arm protruding from the front of the blade attachment (or cuff) and converts the vertical motion of the track rods to a rotation of the blade about the pitch hinge, which is conventionally aligned with the quarter chord of the blade. On a main rotor blade, the pitch horn is a long component as it extends radially inward from the cuff to connect with the track rod on the axis of the flap hinge of the particular blade. In that way any flapping motion of the blade does not cause any (or very little) motion of the bearing connecting the pitch horn to the track rod and will thus remove any tendency of the blade pitch to change as the blade flaps (see Fig. 6.11), in other words there is no *pitch–flap coupling*.

Tail rotors, however, have no need for the disc attitude to be controlled and so do not require cyclic pitch. They are only required to generate thrust and consequently to provide a torque about the main rotor shaft to balance the main rotor torque and to control the helicopter in yaw. In order to obtain the lightest tail rotor design, lag hinges are usually not included in the tail rotor with the proviso that any flapping caused by disturbances is kept to an absolute minimum if not totally avoided. Blade flapping can be triggered by the effects of forward flight or by aerodynamic disturbances and a means of minimising this is included in a tail rotor control system and is known as a $\delta_3$ hinge (see Plate 6.4). The blade flapping motion is governed by the balance between aerodynamic and centrifugal forces, and so if a blade is triggered to flap in a particular direction then a pitch angle change in opposition will minimise the blade flapping motion. This is achieved passively by arranging the pitch horn/track rod connection to lie off the flap hinge line. In other words, pitch flap coupling is encouraged. This is shown in Fig. 6.12. The absence of lag hinges on a tail rotor hub also brings the benefit of this rotor avoiding a ground resonance instability.

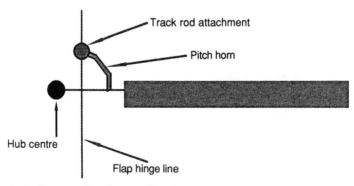

**Fig. 6.11.** Pitch–flap coupling for a main rotor.

**Plate 6.4.** The tail rotor control system of a Sikorsky S-65A. Note the absence of lag hinges and the inclusion of the $\delta_3$ hinge.

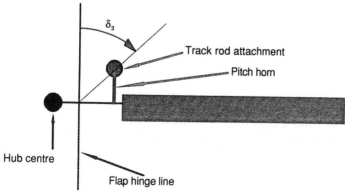

**Fig. 6.12.** Pitch–flap coupling for a tail rotor.

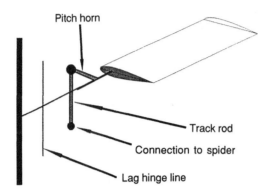

**Fig. 6.13.** Pitch–lag coupling.

## Pitch–lag coupling

Although the track rods impart pitch angle changes by a movement parallel to the shaft because they are mechanical links between the spider arms and pitch horns, they will alter their alignment with the axis of the rotor shaft as they move. Since they are not always exactly perpendicular to the pitch horn over the whole pitch range, any swinging of the blades in lag will cause small changes in pitch angle (see Fig. 6.13). The phenomenon appears to the pilot under high rotor loading conditions as a stirring motion. It usually indicates that the dampers fitted to the blade lag hinges are not as effective as when new and are the first indications of a possible need for replacement.

## Semi-rigid rotors

In these rotors the mechanical complexity of the three hinge arrangement as used in a fully articulated rotor is avoided by using flexible elements in the rotor hub design (see Plate 6.1). They have the advantage of aerodynamic cleanliness compared with a typical articulated rotor and will give the helicopter a greater degree of control, making for a more manoeuvrable aircraft. A further advantage is that they require less maintenance. However, they can have a disadvantage in the form of a greater tendency to transmit vibration from hub to fuselage. In typical designs like the MBB Bolkow 105 (Fig. 6.14) or the Westland Lynx (Fig. 6.15), it is not possible to separate flap and lag hinges totally. This is because some of the flexures bend in both freedoms. The BO105 has the pitch hinge furthest inboard, followed by an element which allows for both flap and lag motions. The Lynx, however, has a flap flexure furthest inboard, followed by a pitch hinge, followed by a flexure which is predominantly lag, but also provides a proportion of the flap deflection. This flexible element, the so-called "dogbone", is situated outside of the pitch hinge position and this brings in to play the phenomenon of *pitch–flap–lag coupling*. With reference to Fig. 6.16, when this flexure is deflected in a flapwise sense, a lagwise force generates a moment about the pitch hinge. A

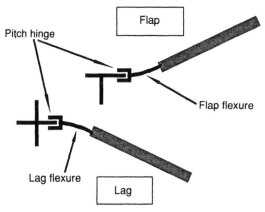

**Fig. 6.14.** Pitch, flap or lag, Bolkow 105.

moment is also generated when a lagwise deflection of the flexure is accompanied by a flapwise force. However, Fig. 6.16 shows that these two pitching moments, with a common convention for positive flapwise and lagwise senses, act in opposite directions. This fact allows pitch–flap–lag coupling to be minimised. It is achieved by making the bending stiffness of the flexure equal in both flapwise and lagwise directions: the so-called "matched stiffness". For this reason the dogbone has a circular cross-section.

## Elastomeric hinges

A recent development is the elastomeric hinge which combines flapping, lagging, and pitching degrees of freedom in one unit and hence at one location. Figure 6.17 shows a schematic diagram of how blocks of an elastomeric compound can allow the blade its three degrees of freedom. The elastomeric bearing consists of

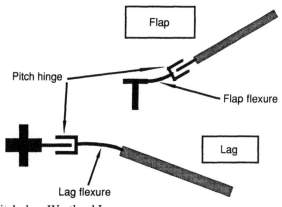

**Fig. 6.15.** Flap, pitch–lag, Westland Lynx.

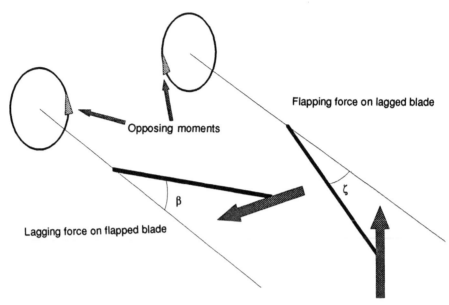

Flapping force on lagged blade

Opposing moments

$\beta$

$\zeta$

Lagging force on flapped blade

**Fig. 6.16.** Pitch–flap–lag coupling.

concentric laminations of elastomer and metal bonded together in alternate layers. The centrifugal force is taken by the elastomer in compression, whilst pitching, lagging and flapping motions are achieved by shear in the elastomer layers. Such a hub may obviate some of the problems of mechanical hinges such as mechanical complexity, maintenance and reliability.

In recent times many different types of main rotor hinge design have been used and Prouty[14] describes a selection of these.

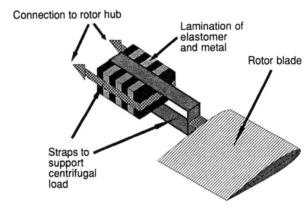

Connection to rotor hub

Lamination of
elastomer
and metal

Rotor blade

Straps to
support
centrifugal
load

**Fig. 6.17.** Elastomeric hinge.

# Blade flapping equation of motion

To determine the flapping behaviour of a rotor blade, the following analysis examines the various forcings on the blade. It concentrates on the aerodynamic forces driving the blade out of the plane of rotation, and the considerable centrifugal forces acting to restore the blade to the plane of rotation. The blade is assumed to be a rigid body freely hinged at the hub. (In reality a rotor blade is flexible, particularly in flap, so due account must be taken of this fact. Chapter 8 describes how the flexibility of a rotor blade can be modelled.) The simple articulated rotor dynamic model consists of the freely hinged blade which is rigid in bending. The semi-rigid rotor blade can be modelled in a similar way, however the bending moment in the flexible elements must be catered for. This can be accomplished by adding a torsional spring (placed at the flapping hinge) to the articulated rotor blade model. As will be shown, the effect of this spring is best described using the natural frequency of the flapping blade.

The lagging behaviour of the rotor blade can be analysed in a similar manner. Extension to a blade which is semi-rigid in lag can also be achieved by including a torsional spring, aligned to provide stiffness in the lag direction. With reference to Fig. 6.18, consider a small blade element, at radius $r$, and possessing a mass $m \cdot dr$, where $m$ is the mass per unit length of the blade and is assumed to be constant. The blade is inclined by an angle $\beta$ (assumed to be small) to the rotor plane. The flapping hinge is located off the shaft centre by an amount $e_f \cdot R$, the flap hinge offset. The centrifugal force is then given by

$$dCF = m\,dr\,\Omega^2 r \qquad (6.1)$$

Taking moments about the flapping hinge, the equation of motion becomes

$$I_\beta \ddot{\beta} = \int_{e_f R}^{R} dL(r - e_f R)$$

$$- \int_{e_f R}^{R} dCF(r - e_f R)\beta \qquad (6.2)$$

$$- k_\beta \beta$$

(Due to high centrifugal force loading (CF), gravitational effects are neglected.)

Note that the small angle approximation of $\beta \approx \sin\beta$ is used, and the blade aerofoil section is assumed to extend from the flapping hinge offset to the blade tip. The first term in Eqn (6.2) is the aerodynamic driving force, the second term the centrifugal stiffening and the third is the restoring effect of a spring placed at the flapping hinge. As previously mentioned, the provision of a spring allows a semi rigid rotor to be modelled, albeit in a simplified manner, because of the rigid blade assumption.

Analysing each moment term we have the following equations.

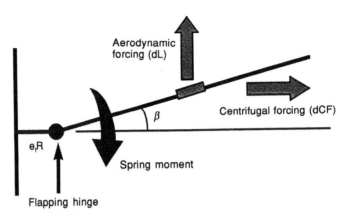

**Fig. 6.18.** Blade forces in flapping motion.

Aerodynamic

$$\int_{e_f R}^{R} dL(r - e_f R) = M_A \tag{6.3}$$

Centrifugal stiffness

$$-\int_{e_f R}^{R} m\Omega^2 r(r - e_f R)\beta\, dr = m\Omega^2 \beta \int_{e_f R}^{R} r(r - e_f R)\, dr$$

$$= m\Omega^2 \beta \left[\tfrac{1}{3} - \tfrac{1}{2} e_f\right] R^3 + O(e_f^2) \tag{6.4}$$

Spring stiffness

$$-k_\beta \beta \tag{6.5}$$

Now the flapping inertia of the blade about the flapping hinge is given by

$$I_\beta = \int_{e_f R}^{R} (r - e_f R)^2 m\, dr$$

$$= mR^3\left(\tfrac{1}{3} - e_f\right) + O(e_f^2) \tag{6.6}$$

The flapping hinge offset, $e_f$, which is expressed relative to the rotor radius, is of the order of 0.05 and it is therefore reasonable to derive the flapping equation to first order terms in $e_f$ only.

Assembling the equation of motion gives

$$mR^3\left(\tfrac{1}{3} - e_f\right)\ddot{\beta} = M_A - m\Omega^2 \beta R^3\left(\tfrac{1}{3} - \tfrac{1}{2} e_f\right) - k_\beta \beta \tag{6.7}$$

from which

$$\ddot{\beta} + \Omega^2 \left[ \frac{k_\beta}{\Omega^2 I_\beta} + 1 + \frac{3e_f}{2(1 - 3e_f)} \right] \beta = \frac{M_A}{I_\beta} \tag{6.8}$$

In order to examine each of the terms in Eqn (6.8), several specific cases are discussed below.

If the rotor is *in vacuo*, we have

$$M_A = 0 \tag{6.9}$$

Also for an articulated rotor

$$k_\beta = 0 \tag{6.10}$$

With these simplifications Eqn (6.8) reduces to the form

$$\ddot{\beta} + \omega_\beta^2 \beta = 0 \tag{6.11}$$

This corresponds to purely the dynamic behaviour of an articulated rotor, *in vacuo*, which is simple harmonic motion with frequency $\omega_\beta$, where

$$\omega_\beta^2 = \Omega^2 \left( 1 + \frac{3e_f}{2(1 - 3e_f)} \right) \approx \Omega^2 \left( 1 + \frac{3}{2} e_f \right) \tag{6.12}$$

Often the frequency $\omega_\beta$ is expressed relative to the rotational speed $\Omega$ via

$$\lambda_\beta^2 = \left[ \frac{\omega_\beta}{\Omega} \right]^2 = 1 + \frac{3}{2} e_f \tag{6.13}$$

It is worth noting that $\lambda_\beta$ is equal to unity when $e_f = 0$, i.e. a teetering or see-saw rotor.

In order to study a semi-rigid rotor, the effect of the spring needs to be isolated which is best performed by considering a stationary rotor. With a spring restraint, and for a stopped rotor, the non-dimensional flapping frequency is defined by

$$\lambda_{\beta_0}^2 = \frac{k_\beta}{I_\beta \Omega^2} \tag{6.14}$$

This is the natural flapping frequency of a stationary rotor blade relative to the normal operating speed of the rotor.

In order to compare this frequency with the results of Eqn (6.13), a normalising with respect to the normal rotor speed, $\Omega$, can be performed. Equation (6.14) expresses the flapping frequency for a stopped rotor, but divided by the normal rotor speed from which we obtain for a general rotor *in vacuo*

$$\lambda_\beta^2 = \left[1 + \tfrac{3}{2}e_f\right] + \left[\lambda_{\beta_0}^2\right] \tag{6.15}$$

The flapping frequency can be seen to originate from two sources. Firstly the contribution of centrifugal forces, and secondly, with a semi-rigid rotor, the elastic bending stiffness afforded by the rotor hub.

## Control response in hover

Having established the basic form of the flapping equation, it is now used to determine the flapping motion of a rotor in hover. For this initial look at solving the flapping equation, the flapping hinge offset is taken to be zero. So far the aerodynamic moment has not been investigated and this is the first to be determined. The incidence at a general blade station is the difference between the blade geometric pitch angle, $\theta$, and the inflow angle $\phi$. The blade is assumed to be untwisted and so, for this example, $\theta$ is not a function of rotor radius $r$. The inflow angle, $\phi$, occurs because of the downflow relative to the blade $U_p$, and the tangential velocity $U_T$ due to the rotor rotation. $U_p$ consists of the rotor downwash $v_i$, assumed constant, and that downflow induced by the blade flapping motion. This is a rotation angular velocity of $\dot\beta$ which at a rotor radius $r$ gives a downflow velocity of $\dot\beta r$.

With reference to the force diagram (Fig. 6.18)

$$U_p = \dot\beta r + v_i$$
$$U_T = \Omega r \tag{6.16}$$

Aerodynamic force

$$\frac{1}{2}\rho V_T^2 x^2 cR\,\mathrm{d}xa\left(\theta - \frac{\dot\beta r}{\Omega r} - \frac{v_i}{\Omega r}\right) \tag{6.17}$$

Centrifugal force

$$\Omega^2 r\,m\,\mathrm{d}r \tag{6.18}$$

Aerodynamic force flapping moment

$$M_A = \frac{1}{2}\rho V_T^2 cRaR\int_0^1 x^3\left(\theta - \frac{\dot\beta}{\Omega} - \frac{\lambda_i}{x}\right)\mathrm{d}x$$

$$= \frac{1}{8}\rho acR^4\Omega^2\left(\theta - \frac{\dot\beta}{\Omega} - \frac{4\lambda_i}{3}\right) \tag{6.19}$$

Centrifugal force flapping moment

$$M_{CF} = -\int_0^R m\,dr\,\Omega^2 rr\beta$$
$$= -\Omega^2 \beta I_\beta \qquad\qquad (6.20)$$

where $I_\beta$, the flapping inertia of the blade, is given by

$$I_\beta = \int_0^R mr^2\,dr \qquad\qquad (6.21)$$

The equation of motion becomes

$$I_\beta \ddot{\beta} = -\Omega^2 \beta I_\beta + \frac{1}{8}\rho a c R^4 \Omega^2 \left(\theta - \frac{\dot{\beta}}{\Omega} - \frac{4\lambda_i}{3}\right) \qquad\qquad (6.22)$$

which on rearrangement gives

$$\ddot{\beta} + \Omega^2 \beta = \frac{\rho a c R^4}{I_\beta} \cdot \frac{1}{8}\left(\theta - \frac{\dot{\beta}}{\Omega} - \frac{4\lambda_i}{3}\right)\Omega^2 \qquad\qquad (6.23)$$

Now the quantity

$$\gamma = \frac{\rho a c R^4}{I_\beta} \sim \frac{\text{aerodynamic forces}}{\text{inertial forces}} \qquad\qquad (6.24)$$

is called the *Lock number* of the blade.

The importance of the Lock number lies in the fact that when the flapping equations of motion are normalised, it is the factor which connects the aerodynamic forces and the dynamic behaviour of the rotor blades.

Now the derivatives of the flapping angle $\beta$ in (6.23), are with respect to time. For algebraic simplification, it is better to express the blade flapping angle and its appropriate derivatives with respect to the azimuth angle $\psi$. For instance, for time derivatives

$$\dot{\beta} = \frac{d\beta}{dt} \qquad\qquad (6.25)$$

$$\ddot{\beta} = \frac{d^2\beta}{dt^2} \qquad\qquad (6.26)$$

and for azimuthal derivatives

$$\beta' = \frac{d\beta}{d\psi} = \frac{\dot{\beta}}{\Omega} \tag{6.27}$$

$$\beta'' = \frac{d^2\beta}{d\psi^2} = \frac{\ddot{\beta}}{\Omega^2} \tag{6.28}$$

The conversions between timewise and azimuthal derivatives being a result of

$$\psi = \Omega t \tag{6.29}$$

from which, after simplification, the equation of motion (6.23) becomes

$$\beta'' + \frac{\gamma}{8}\beta' + \beta = \frac{\gamma}{8}\left(\theta - \frac{4\lambda_i}{3}\right) \tag{6.30}$$

This is now the usual form for a forced simple harmonic oscillation (mass–spring–damper), where the mass term represents the flapping inertia, the spring term the centrifugal stiffening, and the damping comes from the aerodynamics (via the Lock number).

The input to the rotor is via collective and cyclic pitch angles which defines the blade pitch angle by

$$\theta = \theta_0 - A_1 \cos\psi - B_1 \sin\psi \tag{6.31}$$

The blade flapping motion is expressed in a similar form

$$\beta = a_0 - a_1 \cos\psi - b_1 \sin\psi \tag{6.32}$$

where $a_0$ is the constant blade flapping angle known as the coning, $a_1$ is the rearward tilt of the rotor disc plane, and $b_1$ the lateral tilt, advancing side down. Substitution of Eqns (6.31) and (6.32) into (6.30) and equating coefficients of 1, $\cos\psi$ and $\sin\psi$ yields the following results.

Constant terms

$$a_0 = \frac{\gamma}{8}\left(\theta_0 - \frac{4\lambda_i}{3}\right) \tag{6.33}$$

$\sin\psi$ terms

$$a_1 = -B_1 \tag{6.34}$$

$\cos\psi$ terms

$$b_1 = A_1 \tag{6.35}$$

The coning angle, $a_0$, is dependent on the Lock number, collective pitch and rotor downwash. The disc tilt angle expressions (6.34) and (6.35) show how the angles are equal to a cyclic pitch angle, with a delay of exactly 90°. For an articulated rotor with zero flapping hinge offset a change to the blade pitch angle results in the maximum flapping excursion after 90° of rotor rotation. Now a cyclic pitch input occurs at a frequency of once per rotor revolution and seeing as the rotor blade also has a flapping frequency of the same value $(\lambda_\beta = 1)$ the blade flapping is in resonance with the forcing, hence the 90° phase lag is consistent with standard results obtained from dynamic equations of the form of (6.30).

## Flapping equation in hover $(e_f \neq 0)$

The analysis of blade flapping can be extended by allowing the flapping hinge offset to be non-zero. As already shown, this increases the natural frequency above the rotor speed and the relationship between cyclic pitch and blade flapping becomes more complicated. This also has an effect on the azimuthal angle between maximum cyclic pitch and the corresponding maximum flapping angle excursion. This is not surprising since in this case the cyclic pitch still forces the blade at once per revolution, but with an increase in natural frequency, we now have a forced system with a situation below resonance with damping (from the aerodynamics) which will have a phase delay of less than 90°.

Returning to the basic flapping equation

$$\ddot{\beta} + \Omega^2 \left( \frac{k_\beta}{\Omega^2 I_\beta} + 1 + \frac{3}{2} e_f \right) \beta = \frac{M_A}{I_\beta} \tag{6.36}$$

which for convenience can be rewritten as

$$\ddot{\beta} + \Omega^2 \lambda_\beta^2 \beta = \frac{M_A}{I_\beta} \tag{6.37}$$

and again assuming the blade aerofoil section spans the blade from the flapping hinge to the tip, we have for the aerodynamic flapping moment

$$M_A = \int_{e_f R}^{R} dL \cdot (r - e_f R) \tag{6.38}$$

But

$$dL = \frac{1}{2} \rho \Omega^2 r^2 c \, dr a \left( \theta - \frac{\dot{\beta}(r - e_f R)}{\Omega r} \right)$$

$$= \frac{1}{2} \rho \Omega^2 R^2 cRa \left( \theta - \frac{\dot{\beta}(x - e_f)}{\Omega x} \right) x^2 \, dx \tag{6.39}$$

Hence, on collecting these results and non-dimensionalising the integral, the aerodynamic moment is given by

$$M_A = \frac{1}{2}\rho\Omega^2 R^4 ca \int_{e_f}^{1}\left[\theta - \frac{\dot{\beta}}{\Omega}\frac{(x-e_f)}{x}\right]x^2(x-e_f)\,dx$$

$$= \frac{1}{2}\rho acR^4\Omega^2\left[\theta\int_{e_f}^{1}x^2(x-e_f)\,dx - \frac{\dot{\beta}}{\Omega}\int_{e_f}^{1}x(x-e_f)^2\,dx\right] \tag{6.40}$$

from which

$$\frac{M_A}{I_\beta} = \frac{1}{2}\Omega^2\gamma\left[\theta\int_{e_f}^{1}(x^3 - x^2 e_f)\,dx - \frac{\dot{\beta}}{\Omega}\int_{e_f}^{1}(x^3 - 2x^2 e_f + e_f^2 x)\,dx\right]$$

$$= \Omega^2\frac{1}{2}\gamma\left[\theta\left(\frac{1}{4} - \frac{e_f}{3}\right) - \frac{\dot{\beta}}{\Omega}\left(\frac{1}{4} - \frac{2}{3}e_f\right)\right] + O(e_f^2) \tag{6.41}$$

Substituting this result in the flapping equation gives

$$\ddot{\beta} + \Omega^2\lambda_\beta^2\beta = \frac{\gamma}{8}\Omega^2\left[\theta\left(1 - \frac{4}{3}e_f\right) - \frac{\dot{\beta}}{\Omega}\left(1 - \frac{8}{3}e_f\right)\right] \tag{6.42}$$

Transforming the independent variable from time to azimuth angle $\psi$ and simplifying gives

$$\beta'' + \frac{\gamma}{8}\beta'\left(1 - \frac{8}{3}e_f\right) + \lambda_\beta^2\beta = \frac{\gamma}{8}\left(1 - \frac{4}{3}e_f\right)\theta \tag{6.43}$$

Using the definitions in (6.31) and (6.32), Eqn (6.43) becomes

$$[a_1\cos\psi + b_1\sin\psi] + \frac{\gamma}{8}\left(1 - \frac{8}{3}e_f\right)[a_1\sin\psi - b_1\cos\psi]$$

$$+ \lambda_\beta^2[a_0 - a_1\cos\psi - b_1\sin\psi]$$

$$= \frac{\gamma}{8}\left(1 - \frac{4}{3}e_f\right)[\theta_0 - A_1\cos\psi - B_1\sin\psi] \tag{6.44}$$

Equating like terms gives:

   constant terms

$$\lambda_\beta^2 a_0 = \frac{\gamma}{8}\left(1 - \frac{4}{3}e_f\right)\theta_0 \tag{6.45}$$

cos $\psi$ terms

$$a_1(1-\lambda_\beta^2)-\frac{\gamma}{8}\left(1-\frac{8}{3}e_f\right)b_1 = \frac{\gamma}{8}\left(1-\frac{4}{3}e_f\right)(-A_1)$$  (6.46)

sin $\psi$ terms

$$a_1\frac{\gamma}{8}\left(1-\frac{8}{3}e_f\right)+(1-\lambda_\beta^2)b_1 = \frac{\gamma}{8}\left(1-\frac{4}{3}e_f\right)(-B_1)$$  (6.47)

Or in matrix form

$$\begin{bmatrix} (1-\lambda_\beta^2) & -\frac{\gamma}{8}\left(1-\frac{8}{3}e_f\right) \\ \frac{\gamma}{8}\left(1-\frac{8}{3}e_f\right) & (1-\lambda_\beta^2) \end{bmatrix}\begin{bmatrix} a_1 \\ b_1 \end{bmatrix} = \begin{bmatrix} -\frac{\gamma}{8}\left(1-\frac{4}{3}e_f\right)A_1 \\ -\frac{\gamma}{8}\left(1-\frac{4}{3}e_f\right)B_1 \end{bmatrix}$$  (6.48)

which has the solution

$$\begin{bmatrix} a_1 \\ b_1 \end{bmatrix} = \frac{-\frac{\gamma}{8}\left(1-\frac{4e_f}{3}\right)}{(1-\lambda_\beta^2)^2 + \frac{\gamma^2}{64}\left(1-\frac{8e_f}{3}\right)^2}\begin{bmatrix} (1-\lambda_\beta^2) & \frac{\gamma}{8}\left(1-\frac{8}{3}e_f\right) \\ -\frac{\gamma}{8}\left(1-\frac{8}{3}e_f\right) & (1-\lambda_\beta^2) \end{bmatrix}\begin{bmatrix} A_1 \\ B_1 \end{bmatrix}$$  (6.49)

The situation is therefore not as clear cut as the previous situation of $e_f = 0$, and the following analysis shows the effect on phase delay. By differentiating Eqns (6.31) and (6.32) the azimuth angles for maximum cyclic pitch $\psi_C$ and maximum flapping $\psi_F$ are given by

$$\tan\psi_C = \frac{B_1}{A_1} \quad \text{and} \quad \tan\psi_F = \frac{b_1}{a_1}$$  (6.50)

If the phase delay of the flapping behind the cyclic pitch is $\phi_F$ then

$$\phi_F = \psi_F - \psi_C$$  (6.51)

therefore

$$\tan\phi_F = \frac{\tan\psi_F - \tan\psi_C}{1 + \tan\psi_F \tan\psi_C} = \frac{\dfrac{b_1}{a_1} - \dfrac{B_1}{A_1}}{1 + \dfrac{b_1}{a_1}\cdot\dfrac{B_1}{A_1}}$$  (6.52)

Substituting (6.49) into (6.52) and simplifying gives the following expression for the phase delay

$$\tan \phi_F = \frac{\gamma\left(1 - \dfrac{8e_f}{3}\right)}{8(\lambda_\beta^2 - 1)} \tag{6.53}$$

The effect of the flapping hinge offset, $e_f$, on flapping frequency and phase delay is shown in Fig. 6.19 for a Lock number of 8.

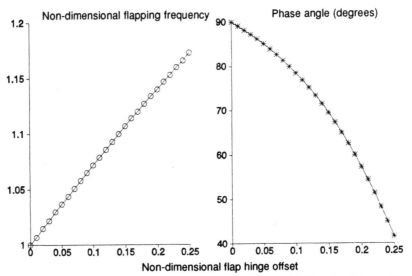

**Fig. 6.19.** Variation of flapping frequency and phase delay with flapping hinge offset.

## The effect of a $\delta_3$ hinge

In addition to the effects of blade flapping motion, changes to incidence angles can occur if pitch flap coupling is introduced via the $\delta_3$ hinge. This is pertinent to the tail rotor. The change comes from the alterations to the geometric pitch angle through the mechanical linkages of the control system, and in particular to the location of the end of the pitch horn relative to the flapping hinge line. With reference to Fig. 6.12, if the flapping excursions are small, then the change of pitch angle will vary linearly with flapping angle, the constant given by $\tan \delta_3$. If, for instance, the blade flaps up by an amount $\beta$, with the pitch horn–track rod connection point remaining fixed, the pitch angle will be reduced by an angle

$$\beta \cdot \tan \delta_3 \tag{6.54}$$

i.e. the pitch angle term in the right-hand side of the flapping equation (6.30) becomes

$$\theta - \beta \cdot \tan \delta_3 \quad \text{or} \quad \theta - \beta \cdot \frac{\partial \theta}{\partial \beta} \tag{6.55}$$

Therefore the basic flapping equation now becomes with the addition of the $\delta_3$ hinge

$$\beta'' + \frac{\gamma}{8} \beta' + \left(1 + \frac{\gamma}{8} \tan \delta_3\right) \beta = \frac{\gamma}{8}\left(\theta - \frac{4}{3} \lambda_i\right) \tag{6.56}$$

The effect of the $\delta_3$ hinge is placed in the spring term of the left-hand side of (6.56) the solution of which gives the following conclusions

(a)  The coning angle is reduced for a given collective pitch angle

$$a_0 = \frac{\gamma\left(\theta_0 - \frac{4}{3} \lambda_i\right)}{8 + \gamma \tan \delta_3} \tag{6.57}$$

(b)  The flapping frequency has been increased by the aerodynamic coupling to

$$\omega_\beta = \Omega \sqrt{1 + \frac{\gamma}{8} \tan \delta_3} \tag{6.58}$$

The aerodynamics of the blade influences the flap damping of any rotor, but pitch flap coupling also allows it to affect the frequency.

A fuller analysis of $\delta_3$ and its effect is presented in Chapter 11.

## General flapping motion

As already shown, the flapping equation is a differential equation in $\beta$, because each of the moments, including the aerodynamic, are functions of $\beta$ and its timewise derivative $\dot{\beta}$. The aerodynamic moment produces both a damping moment and a forcing function. So far the discussion has been concerned with hover where axial symmetry of the inflow occurs. However, as the helicopter leaves the hovering flight state and enters translational flight there is now a periodicity introduced to the aerodynamic excitation caused by the different flow conditions on the advancing and retreating sides of the rotor disc originating in the summation of the rotational and forward velocities of the rotor blades. If this forward flight situation is examined without application of cyclic pitch control, the aerodynamic forces are approximately proportional to the square of the velocity at the blade, namely

$$(x + \mu_x \sin \psi)^2 = x^2 + 2x\mu_x \sin \psi + \mu_x^2 \sin^2 \psi$$

$$= \left(x^2 + \frac{\mu_x^2}{2}\right) + (2x\mu_x)\sin \psi - \left(\frac{\mu_x^2}{2}\right)\cos 2\psi \qquad (6.59)$$

i.e. a first harmonic change in velocity, when squared imposes a constant forcing plus a first and second harmonic disturbance.

Up until now only constant and first harmonic flapping have been identified and the three flapping coefficients $a_0$, $a_1$ and $b_1$ have been determined by the equations formed by equating the coefficients of 1, $\cos \psi$ and $\sin \psi$. The other terms, such as $\sin 2\psi$ in hover have not occurred, but to introduce them through the effects of forward flight will require additional terms in the flapping motion expression to restore balance to the solution.

If $\cos 2\psi$ and $\sin 2\psi$ terms are permitted in the aerodynamic forcing as shown in (6.59), then the flapping response must be of the form

$$\beta = a_0 - a_1 \cos \psi - b_1 \sin \psi - a_2 \cos 2\psi - b_2 \sin 2\psi \qquad (6.60)$$

In hover and axial flight, the response is of a constant coefficient $a_0$, which is found by equating the centrifugal force and lift moments. In forward flight, the disc attitude will now tilt away from the plane perpendicular to the rotor shaft with $a_1$ representing the amplitude of pure cosine motion and therefore the rearward disc tilt angle (see Fig. 6.20). The term $b_1$ represents the amplitude of pure sine flapping motion, and hence represents lateral disc tilt advancing side down (see Fig. 6.21). In extending the solution, the terms $a_2$ and $b_2$ occur, which are second harmonic amplitudes representing a weaving in and out of the disc plane. Typically, $b_1 \approx a_1$ and $a_2$, $b_2 \approx 10^{-1} a_1$, $b_1$.

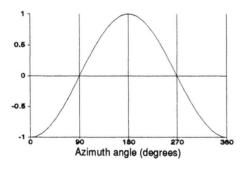

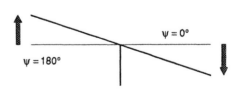

**Fig. 6.20.** $a_1$ Flapping motion.

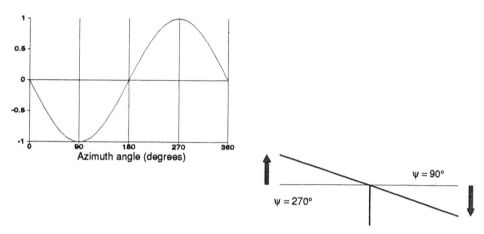

**Fig. 6.21.** $b_1$ Flapping motion.

The extension to second harmonic flapping behaviour can, in theory, be extended to higher harmonics allowing more general flapping motion to be defined. Whilst this has the appeal of completeness it must be remembered that this is a rigid blade model which will give a false notion of blade behaviour at frequencies that are well in excess of the rotor speed. In reality, this type of behaviour will be much affected by elastic deformations of the blade making the rigid blade model only of use for low harmonic behaviour. The modelling of elastic blade behaviour is discussed in Chapter 8.

## Reference planes

The flapping motion of the rotor blades can be referenced to any axis system, however three different frames of reference, or planes of rotation, have evolved over the years, each having its own advantages. (Feathering is the alternative term for pitching.)

The *plane of no feathering* is a plane within which an observer will see no cyclic feathering of the blades, only a constant pitch angle. Figure 6.22 shows its existence. Two blade sections are shown which represent a chordwise section at a given radial position of a blade when on opposite sides of the rotor disc, i.e. when the azimuth angles are 90° and 270°. Each section will have a reference line and the figure also shows axes which are aligned with the bisectors of the section reference line at the two blade azimuth positions. An observer in these axes will see no change in blade pitch (or feathering). A change in collective pitch will alter the inclination of the section lines to the axes, but this change will remain constant to the observer as the blades rotate about the shaft. Blade flapping only will be observed.

The *plane of no flapping* is represented by the plane traced out by the rotor tips. An observer in this plane will see the blades set at constant flapping angles, i.e. steady coning. However, cyclic feathering of the blades will be observed.

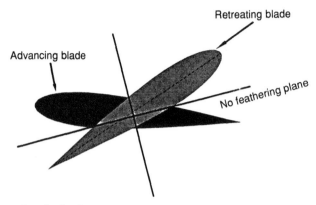

**Fig. 6.22.** Plane of no feathering.

The *plane normal to the shaft axis* is the third and most complicated. In this plane of reference an observer will see both cyclic feathering and flapping. However, it has the distinct advantage of being firmly linked to a physical part of the rotor system, i.e. the rotor shaft. For this reason the shaft axis is commonly used in industry, and is used throughout this book.

## Equivalence of flapping and feathering

The hovering rotor analysis, through Eqns (6.34) and (6.35), shows a solid link between the applied cyclic pitch angles and the resulting maximum flapping angles. Consider a blade rotating about a vertical shaft with zero pitch angle. Figure 6.23 shows a view from the side of the rotor with the disc tilted (by an angle $\beta_{\text{MAX}}$) to the right with the near blade passing from left to right and vice versa as the blade moves to the opposite side of the disc.

An observer in the *no feathering plane* sees no pitch change, but the blades are flapping with maximum flapping angle of $\beta_{\text{MAX}}$. An observer in the *no flapping (or tip path) plane* sees a constant flapping angle, but the blade pitch is changing with maximum value of $\theta_{\text{MAX}}$. Figure 6.23 shows that $\beta_{\text{MAX}} = \theta_{\text{MAX}}$, hence the equivalence of feathering and flapping. Strictly, this equality is only applicable to a teetering rotor hub, where the flapping hinge offset is zero and the flapping frequency is equal to the rotor speed, however, the link is very strong for all rotors particularly articulated types. It shows geometrically the close links between flapping and feathering blade motion established before in Eqns (6.34) and (6.35).

## Blade lag motion

The necessity of the blade lag degree of freedom has been already discussed. However, to analyse the blade behaviour in the rotor plane, the natural frequency of a rotor blade in lag needs to be determined. The assembly of the equations of motion follows closely to that of the flapping motion. The aerodynamic drag force is omitted

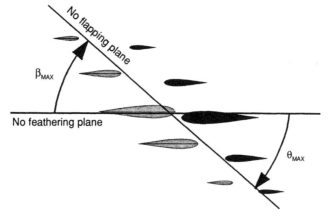

**Fig. 6.23.** Flapping/feathering equivalence.

as was lift in the case of the natural flapping frequency, and the force diagram is shown in Fig. 6.24. The lag angle is usually denoted by $\zeta$.

Taking moments about the lag hinge gives

$$I_\zeta \ddot{\zeta} = -\int_{e_L R}^{R} \mathrm{d}(\mathrm{CF}) \cdot e_L R \sin\theta - k_\zeta \zeta \tag{6.61}$$

where $k_\zeta$ is the lag spring rate.

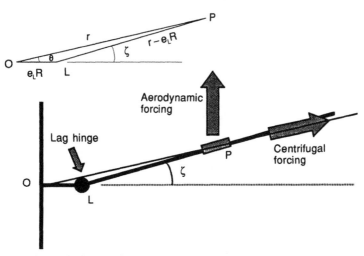

**Fig. 6.24.** Blade forces in lag motion.

Now the moment of inertia of the blade in lag about the lag hinge is given by

$$I_\zeta = \int_{e_L R}^{R} m(r - e_L R)^2 \, dr$$

$$= \tfrac{1}{3} mR^3 (1 - e_L)^3 \tag{6.62}$$

$$= \tfrac{1}{3} mR^3 (1 - 3e_L) + O(e_L^2)$$

(The mass per unit length has been assumed constant). Also

$$d(CF) = m \cdot dr \cdot \Omega^2 r \tag{6.63}$$

Therefore the centrifugal force restoring moment is

$$-\int_{e_L R}^{R} m \, dr \, \Omega^2 r e_L R \sin\theta = -m\Omega^2 \int_{e_L R}^{R} r e_L R \sin\theta \, dr \tag{6.64}$$

By applying the sine rule to triangle $OLP$ (assuming $\zeta$ is small) we have

$$\frac{r - e_L R}{\sin\theta} = \frac{r}{\sin\zeta} \tag{6.65}$$

From which we find

$$r\sin\theta = (r - e_L R)\sin\zeta \approx (r - e_L R)\zeta \tag{6.66}$$

The CF integral becomes

$$-m\Omega^2 e_L R \int_{e_L R}^{R} (r - e_L R)\zeta \, dr = -m\Omega^2 e_L R^3 \zeta [\tfrac{1}{2}(1 - e_L)^2] \tag{6.67}$$

$$= -m\Omega^2 e_L R^3 \zeta (\tfrac{1}{2} - e_L) + O(e_L^2)$$

From which the equation of motion becomes

$$\tfrac{1}{3} mR^3 (1 - 3e_L)\ddot\zeta + m\Omega^2 e_L R^3 \zeta (\tfrac{1}{2} - e_L) + k_\zeta \zeta = 0 \tag{6.68}$$

This is simple harmonic motion with frequency $\omega_\zeta$, where

$$\omega_\zeta^2 = \frac{3\Omega^2 e_L (\tfrac{1}{2} - e_L)}{(1 - 3e_L)} + \frac{k_\zeta}{I_\zeta} \tag{6.69}$$

$$= \tfrac{3}{2} e_L \Omega^2 + \omega_{\zeta 0}^2 + O(e_L^2)$$

where

$$\omega_{\zeta 0}^2 = \frac{k_\zeta}{I_\zeta} \tag{6.70}$$

and is the natural lag frequency of a blade with the rotor stationary. From these results the lag frequency ratio becomes

$$\lambda_\zeta^2 = \tfrac{3}{2}e_L + \lambda_{\zeta 0}^2 \tag{6.71}$$

where

$$\lambda_{\zeta 0} = \frac{\omega_{\zeta 0}}{\Omega} \tag{6.72}$$

and for a pure articulated rotor

$$\lambda_\zeta = \sqrt{\tfrac{3}{2}e_L} \tag{6.73}$$

The blade moment of inertia about the lag hinge, $I_\zeta$, is close to the value about the flapping hinge. Since the restoring moment (due to the centrifugal restoring moment) is much smaller in lag than flap, it can be seen that the natural frequency in lag is much lower than the value in flap. Flapping motion has a natural frequency in excess of the rotor speed ($1.05\Omega - 1.1\Omega$), whilst for articulated rotors the lag frequency has typical values of $0.2\Omega - 0.3\Omega$.

Flapping motion induces a velocity which changes the blade incidence, and hence the blade section lift, in a sense to oppose the motion and thus impart damping. Motion about the lag hinge changes the section drag through the tangential velocity change, which is a much smaller effect than the incidence change of the flapping motion. The in-plane (lag) damping forces are in consequence much smaller and may be inadequate to fully control any instabilities which may occur. The most feared of these is ground resonance, where lateral vibration of the fuselage couples with blade lag motion to produce a potentially catastrophic occurrence. As the damping in lag due to the aerodynamic effects are usually insufficient, mechanical dampers are fitted to rotor hubs to suppress this phenomenon, which is discussed in Chapter 12.

## Hingeless or semi-rigid rotors

This type of rotor permits the flapping and lagging motions of the blade by means of flexible elements, rather than rotations about hinges. This means that the flapping or lagging hinge is not a precise idea as defined by a bearing assembly, as is the case with an articulated rotor. As the blade moves, the bending of the flexures contributes to the stiffness of the blade with the result of increasing the natural frequencies. A full dynamic analysis of such a rotor system would require a return to the original equations and to derive them again, including the above concepts with appropriate modifications to the dynamic terms.

The original equations can be used, however, by making use of the concept of the equivalent flapping/lagging hinge, which is shown in Fig. 6.25. The rotor blade will mainly deflect in the fundamental mode shape. This has the majority of its curvature in the regions of the rotor close to the hub, i.e. the flexures, whilst the outer portion, corresponding to the blade itself, is essentially straight. In other words, the aerodynamic portion of the blade/hub system behaves as if it is rotating about a fixed point, which can be seen in Fig. 6.25 and is located at the point where the projection of the linear part of the blade intersects the undeflected rotor plane. In this way, an articulated blade with the flapping/lagging hinge situated at the equivalent hinge offset can be used to model the blade movement. Having allowed for the physical movement of the blade, the dynamic equations of motion require that the dynamic characteristics of the rotor blade and hub are also correct. This requires that the natural flapping/lagging frequency must agree with the hingeless blade being modelled. The correct natural flapping/lagging frequency can be achieved by adding a spring to the articulated blade in the appropriate direction, artificially raising the blade/hub stiffness to that seen in the original hingeless rotor.

The concept of the additive spring has already been mentioned in the derivation of Eqn (6.2). The additional stiffness afforded by the spring increases the natural flapping frequency ratio, $\lambda_\beta$, to values typically in the range 1.1–1.2.

## Flapping/thrust equations in forward flight

The flapping equation is now applied to the case of a rotor in forward flight. The aerodynamic forcing will now involve the advance ratio, $\mu$, and its components parallel and perpendicular to the rotor disc, $\mu_x$ and $\mu_z$, respectively. In order to simplify the analysis, terms of order $\mu^2$ and above are neglected. This is purely to clarify the analysis and retention of these terms will not alter the general concepts. This action is justified if the advance ratio is of normal values, and the results should be relevant. The lift coefficient of the blade aerofoil section is assumed to vary linearly with incidence and therefore the blade must be considered as unstalled.

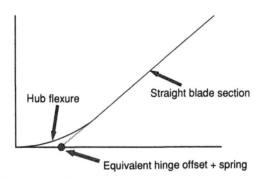

**Fig. 6.25.** Equivalent hinge offset.

In summary, the following assumptions are made for this analysis

(a)  The blade is rigid.
(b)  The flap hinge offset is zero.
(c)  Terms of order $O(\mu^2)$ can be neglected.
(d)  Constant lift curve slope $(a = dC_L / d\alpha)$.

The tangential component, in the plane normal to the rotor shaft axis, of velocity is given by

$$U_T = V_T(x + \mu_x \sin\psi) \tag{6.74}$$

The corresponding component, parallel to the shaft axis, is given by

$$U_P = V_T(\mu_{zD} + \beta'x + \mu_x\beta\cos\psi) \tag{6.75}$$

The final term in this expression has not been encountered before and is due to a component of forward speed $(\mu_x)$ existing in the direction normal to the rotor plane caused by the flapping angle, $\beta$, on the blade. For an upward flapping angle, $(\beta > 0)$, it will vary from the downwash direction at the rear of the disc to upwash at the front. This is catered for by the $\cos\psi$ term. Note that $\mu_x$ and $\mu_z$ are in *shaft axes*.

The pitch, inflow, and incidence angles are therefore

$$\theta = \theta_0 - A_1\cos\psi - B_1\sin\psi$$

$$\phi = \frac{U_P}{U_T} \tag{6.76}$$

$$\alpha = \theta - \phi$$

As before, the flapping angle, and its first two azimuthwise derivatives are

$$\beta = a_0 - a_1\cos\psi - b_1\sin\psi$$

$$\beta' = a_1\sin\psi - b_1\cos\psi \tag{6.77}$$

$$\beta'' = a_1\cos\psi + b_1\sin\psi$$

The aerodynamic contribution to the flapping moment now becomes

$$M_A = \int_0^1 \frac{1}{2}\rho cR\,dx\,U_T^2\,\alpha\,Rx\,a$$

$$= \int_0^1 \frac{1}{2}\rho U_T^2 cR^2 a\left(\theta - \frac{U_P}{U_T}\right)x\,dx \tag{6.78}$$

$$= \frac{1}{2}\rho caR^2\int_0^1 [\theta x U_T^2 - x U_P U_T]dx$$

On substituting Eqns (6.74) and (6.75) into Eqn (6.78), we have

$$
M_A = \frac{1}{2}\rho a c R^2 V_T^2 \left[ \theta \int_0^1 x(x + \mu_x \sin\psi)^2 \, dx \right.
$$

$$
\left. - \int_0^1 x(x + \mu_x \sin\psi)(\mu_{zD} + \beta'x + \mu_x\beta\cos\psi) \, dx \right]
$$

(6.79)

which reduces to

$$
M_A = \tfrac{1}{2} I_\beta \gamma \Omega^2 [\theta I_1 - I_2]
$$

(6.80)

where

$$
I_1 = \int_0^1 (x + \mu_x \sin\psi)^2 x \, dx
$$

$$
= \frac{1}{4} + \frac{2}{3}\mu_x \sin\psi + O(\mu^2)
$$

$$
I_2 = \int_0^1 x(x + \mu_x \sin\psi)(\mu_{zD} + \mu_x\beta\cos\psi + \beta'x) \, dx
$$

$$
= \frac{1}{4}\beta' + \frac{1}{3}(\mu_{zD} + \mu_x\beta\cos\psi + \mu_x\beta'\sin\psi) + O(\mu^2)
$$

(6.81)

Substituting for $\theta$ from (6.76) gives

$$
\theta I_1 - I_2 = (\theta_0 - A_1\cos\psi - B_1\sin\psi)(\tfrac{1}{4} + \tfrac{2}{3}\mu_x\sin\psi)
$$

$$
- \tfrac{1}{4}(a_1\sin\psi - b_1\cos\psi)
$$

$$
- \tfrac{1}{3}(\mu_{zD} + \mu_x\cos\psi[a_0 - a_1\cos\psi - b_1\sin\psi] + \mu_x\sin\psi[a_1\sin\psi - b_1\cos\psi])
$$

(6.82)

From which

$$
\theta I_1 - I_2 = \frac{\theta_0}{4} - \frac{\mu_{zD}}{3} + \cos\psi\left[ -\frac{A_1}{4} + \frac{b_1}{4} - \frac{1}{3}\mu_x a_0 \right]
$$

$$
+ \sin\psi\left[ \frac{2}{3}\mu_x\theta_0 - \frac{B_1}{4} - \frac{a_1}{4} \right]
$$

$$
+ \sin\psi\cos\psi\left[ -\frac{2}{3}\mu_x A_1 + \frac{2}{3}\mu_x b_1 \right]
$$

$$
+ (\cos^2\psi - \sin^2\psi)\left[ \frac{\mu_x}{3} a_1 \right]
$$

$$
+ \sin^2\psi\left[ -\frac{2}{3}\mu_x B_1 \right]
$$

(6.83)

Noting that

$$\sin\psi \cdot \cos\psi = \tfrac{1}{2}\sin 2\psi$$

$$\cos^2\psi - \sin^2\psi = \cos 2\psi \qquad (6.84)$$

$$\sin^2\psi = \tfrac{1}{2}(1 - \cos 2\psi)$$

Equation (6.83) can be rearranged in a Fourier series

$$
\begin{aligned}
\theta I_1 - I_2 = 1 &\left[ \frac{1}{4}\theta_0 - \frac{1}{3}\mu_x B_1 - \frac{\mu_{zD}}{3} \right] \\
&+ \cos\psi \left[ \frac{b_1 - A_1}{4} - \frac{\mu_x}{3}a_0 \right] \\
&+ \sin\psi \left[ \frac{2}{3}\mu_x\theta_0 - \frac{a_1 + B_1}{4} \right] \\
&+ \cos 2\psi \left[ \frac{\mu_x}{3}(a_1 + B_1) \right] \\
&+ \sin 2\psi \left[ \frac{\mu_x}{3}(b_1 - A_1) \right]
\end{aligned}
\qquad (6.85)
$$

## Flapping equation

The flapping equation can now be assembled as before

$$I_\beta \ddot{\beta} = M_A - k_\beta \beta \qquad (6.86)$$

Note that the stiffness term has been introduced as a spring. This means that two stiffness effects have been combined, namely the centrifugal force terms and any flexure stiffness to allow for a hingeless rotor hub.

Expressing the stiffness by means of a spring rate is not particularly useful, however, if it is converted to a frequency, then the dynamic parameters of the blade can be readily introduced.

If the natural flapping frequency *in vacuo* is

$$\omega_\beta = \Omega\lambda_\beta \qquad (6.87)$$

we have

$$k_\beta = I_\beta \Omega^2 \lambda_\beta^2 \qquad (6.88)$$

Equation (6.86) now becomes

$$\Omega^2 I_\beta \beta'' + I_\beta \Omega^2 \lambda_\beta^2 \beta = M_A \qquad (6.89)$$

which on simplifying and substituting Eqn (6.80) gives

$$\beta'' + \lambda_\beta^2 \beta = \frac{M_A}{\Omega^2 I_\beta} = \frac{\gamma}{2}[\theta I_1 - I_2] \tag{6.90}$$

On substituting Eqn (6.77) we have for the left-hand side of (6.90)

$$\begin{aligned}
\beta'' + \lambda_\beta^2 \beta &= a_1 \cos\psi + b_1 \sin\psi + \lambda_\beta^2(a_0 - a_1 \cos\psi - b_1 \sin\psi) \\
&= \lambda_\beta^2 a_0 + a_1(1 - \lambda_\beta^2)\cos\psi + b_1(1 - \lambda_\beta^2)\sin\psi
\end{aligned} \tag{6.91}$$

Also for the right-hand side of (6.90)

$$\frac{M_A}{\Omega^2 I_\beta} = \frac{\gamma}{2}\left( \begin{array}{l}
1\left[\dfrac{\theta_0}{4} - \dfrac{1}{3}\mu_x B_1 - \dfrac{\mu_{zD}}{3}\right] \\[2mm]
+\cos\psi\left[\dfrac{b_1 - A_1}{4} - \dfrac{\mu_x}{3}a_0\right] \\[2mm]
+\sin\psi\left[\dfrac{2}{3}\mu_x\theta_0 - \dfrac{a_1 + B_1}{4}\right] \\[2mm]
+\cos 2\psi\left[\dfrac{\mu_x}{3}(a_1 + B_1)\right] \\[2mm]
+\sin 2\psi\left[\dfrac{\mu_x}{3}(b_1 - A_1)\right]
\end{array} \right) \tag{6.92}$$

Equating coefficients of 1, $\cos\psi$ and $\sin\psi$ in the terms of (6.90) gives the following set of equations

$$\begin{array}{l}
\lambda_\beta^2 a_0 \\[2mm]
\dfrac{\gamma}{6}\mu_x a_0 + (1 - \lambda_\beta^2)a_1 - \dfrac{\gamma}{8}b_1 \\[2mm]
\dfrac{\gamma}{8}a_1 \qquad + (1 - \lambda_\beta^2)b_1
\end{array} = \frac{\gamma}{2}\left( \begin{array}{l}
\left[\dfrac{\theta_0}{4} - \dfrac{\mu_x B_1}{3} - \dfrac{\mu_{zD}}{3}\right] \\[2mm]
\left[-\dfrac{A_1}{4}\right] \\[2mm]
\left[\dfrac{2}{3}\mu_x\theta_0 - \dfrac{B_1}{4}\right]
\end{array} \right) \tag{6.93}$$

The unknowns in the equations include the downwash $\lambda_i$, which forms part of the $\mu_{zD}$ term. The downwash value depends on the rotor thrust which is itself dependent on the collective and cyclic pitch angles applied to the rotor blades. In order to make the set of equations consistent, the link between the downwash $\lambda_i$ and the rotor control angles $\theta_0$, $A_1$ and $B_1$ must be accounted for, hence the thrust equation must be included in the set of equations.

## Thrust equation

Using previous results we have for the rotor thrust

$$
\begin{aligned}
T &= \frac{N}{2\pi} \int_0^{2\pi} d\psi \int_0^1 \frac{1}{2} \rho c R \, dx \, U_T^2 a \alpha \\
&= \frac{N}{2\pi} \int_0^{2\pi} d\psi \int_0^1 \frac{1}{2} \rho c a R \left( U_T^2 \theta - U_P U_T \right) dx
\end{aligned}
\tag{6.94}
$$

By a similar analysis to the flapping equations we have for the right-hand side of (6.94)

$$
\int_0^1 \frac{1}{2} \rho \, caR \left( U_T^2\theta - U_P U_T \right) dx = \frac{1}{2}\rho V_T^2 \, caR \left( \begin{array}{l} 1\left[\dfrac{\theta_0}{3} - \dfrac{1}{2}\mu_{zD} - \dfrac{1}{2}\mu_x B_1\right] \\[2ex] + \cos\psi\left[\dfrac{b_1 - A_1}{3} - \dfrac{1}{2}\mu_x a_0\right] \\[2ex] + \sin\psi\left[-\dfrac{a_1 + B_1}{3} + \mu_x\theta_0\right] \\[2ex] + \cos2\psi\left[\dfrac{\mu_x}{2}(B_1 + a_1)\right] \\[2ex] + \sin2\psi\left[\dfrac{\mu_x}{2}(b_1 - A_1)\right] \end{array} \right)
\tag{6.95}
$$

For the blade flapping description, the azimuthal variation must be retained in order to equate the Fourier terms. The rotor thrust, however, is a mean value and hence obtained, as before, by averaging around the rotor disc by integration.

On averaging around the azimuth we have

$$
T = \frac{1}{2} \rho V_T^2 a N c R \left[ \frac{\theta_0}{3} - \frac{1}{2}\mu_x B_1 - \frac{1}{2}\mu_{zD} \right]
\tag{6.96}
$$

or in coefficient form

$$
\frac{C_T}{s} \cdot \frac{1}{a} = \frac{\theta_0}{3} - \frac{1}{2}\mu_x B_1 - \frac{1}{2}\mu_{zD}
\tag{6.97}
$$

The final assembly of the three flapping equations (6.93) and the thrust equation (6.97) gives the following matrix equation

$$
\begin{bmatrix}
\lambda_\beta^2 & 0 & 0 & 0 \\[6pt]
\dfrac{\gamma}{6}\mu_x & 1-\lambda_\beta^2 & -\dfrac{\gamma}{8} & 0 \\[6pt]
0 & \dfrac{\gamma}{8} & 1-\lambda_\beta^2 & 0 \\[6pt]
0 & 0 & 0 & 1
\end{bmatrix}
\cdot
\begin{bmatrix}
a_0 \\[6pt]
a_1 \\[6pt]
b_1 \\[6pt]
\dfrac{C_T}{s}\cdot\dfrac{1}{a}
\end{bmatrix}
=
$$

$$
\begin{bmatrix}
\dfrac{\gamma}{8} & 0 & -\dfrac{\gamma}{6}\mu_x & -\dfrac{\gamma}{6} \\[6pt]
0 & -\dfrac{\gamma}{8} & 0 & 0 \\[6pt]
\dfrac{\gamma}{3}\mu_x & 0 & -\dfrac{\gamma}{8} & 0 \\[6pt]
\dfrac{1}{3} & 0 & -\dfrac{1}{2}\mu_x & -\dfrac{1}{2}
\end{bmatrix}
\cdot
\begin{bmatrix}
\theta_0 \\[6pt]
A_1 \\[6pt]
B_1 \\[6pt]
\mu_z+\lambda_i
\end{bmatrix}
\qquad (6.98)
$$

There are two types of problem to be solved, namely

1.  Given $C_T/sa$, $a_1$, $b_1$, $\lambda_i$ (from $C_T$); solve for $a_0$, $\theta_0$, $A_1$, $B_1$. In this case, the rotor disc attitude and thrust is known and the analysis enables the rotor control angles and coning angle to be determined. This is the easier solution as the disc attitude and rotor thrust are known, requiring the downwash iteration to be performed once only.
2.  Given $\theta_0$, $A_1$, $B_1$; solve for $a_0$, $a_1$, $b_1$, $C_T/sa$, $\lambda_i$. In this case, the rotor control angles are specified, and the analysis fixes the resulting rotor thrust and disc attitude. The $C_T$ vs $\lambda_i$ (downwash) iteration has to be performed more than once, since the disc attitude, and hence $\mu_x$ and $\mu_z$, are unknown. Values of rotor thrust have to be assumed allowing the iteration to take place. The results of the rotor thrust and downwash calculations must be checked to ensure that they match the rotor pitch inputs and the blade behaviour calculated. This procedure must continue until all the rotor parameters, both input and calculated results must match. The non-linear link between $\mu_x$, $\mu_z$, $C_T$ and $\lambda_i$ means that the solution must be achieved via an iterative scheme.

Note that $\mu_x$ and $\mu_z$ are referred to *shaft axes*. The downwash iteration will require the velocities parallel and normal to the disc plane, i.e. referred to *disc axes*.

The set of equations (6.98) will be used in Chapter 9 when a helicopter trim calculation is described.

# Autorotative performance

Should a helicopter suffer a partial or total power failure, the ability to descend under control and, if necessary, to make a controlled landing is an important safety feature. The essence of autorotation is to turn the rotor by means of an external force enabling rotor thrust to be generated even though the engine(s) are not supplying any torque. The rotor thrust can then be used to allow the necessary descent of the helicopter to be under control and to allow a safe touchdown. The external driving torque can be considered from two viewpoints.

Firstly, the actuator disc model for a rotor in vertical descent has been analysed already in Chapter 4 and the total power is the product of the rotor thrust and the total velocity through the rotor disc to which is added the profile power. The total velocity through the disc can become negative (i.e. upward) if the vertical descent rate is sufficiently high and the possibility of zero power exists. To do this the descent rate must overcome the induced velocity which *always* opposes the rotor thrust, and in addition overcome the profile power. The higher the disc loading, the greater the value of induced velocity and hence the faster the helicopter must descend to achieve the zero total power condition. In vertical flight, normal descent is made in the windmill brake state in which the helicopter descends with the rotor operating like a windmill. In forward flight use can be made of the horizontal speed by raising the nose of the helicopter and allowing the resulting upflow to turn the rotor. This chapter, however, considers vertical autorotation only.

Secondly, the driving torque caused by the descent of the rotor can be derived from a blade element approach. To do this, consider first the case of a hovering rotor. The flow conditions for a typical blade section are shown in Fig. 7.1 and from this it can be seen that there is a drag force at this section equal to

$$dL \sin\phi + dD \cos\phi \qquad (7.1)$$

The first term is the rearward tilt of the lift vector which depends on the downflow velocity through the rotor disc, and the second term is the profile drag. This total drag force must be overcome by the torque supplied by the engine(s).

The rotor thrust is given by

$$dL \cos\phi - dD \sin\phi \qquad (7.2)$$

Consider now the failure of the engine. If the pilot takes no action and the controls are frozen in position, the rotor will slow down due to the drag force expressed in

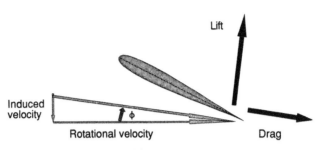

**Fig. 7.1.** Power-on hovering flow condition.

Eqn (7.1). As the rotor slows down, the lift force in Eqn (7.2) will decrease and the helicopter will descend. As time progresses, the situation is reached where the rotor blades will be rotating so slowly that the centrifugal force will be insufficient to maintain them near to the plane of rotation, and excessive flapping motion will take place with the eventual disintegration of the rotor and loss of the aircraft.

In order to save the helicopter, the pilot must therefore take action to reduce the drag (7.1), which is achieved by reducing the inflow angle, $\phi$, to zero by eliminating the lift. This is done by reducing the collective pitch of the rotor blades, reducing the rotor thrust and thereby causing the aircraft to fall under gravity. As the descent progresses air now flows upwards through the rotor so that at some stage the vector diagram looks like Fig. 7.2.

Now that $\phi$ is negative, the contribution to the blade drag due to the lift is opposite in direction to that of the profile drag and it is this induced thrust which provides a means of keeping the rotor rotating. The lift combines with the vertical component of the drag force to produce a rotor thrust force in opposition to the weight.

After the initiation of the helicopter's descent, the pilot adjusts the collective pitch of the blades to obtain a sufficient rotational speed of the helicopter rotor and a lift equal to the aircraft weight. This enables the helicopter to descend steadily at a given rate of descent.

The following analysis uses actuator disc and blade element theory to calculate the descent velocity required to achieve the steady autorotative state. In order to overcome the drag on the rotor blades, energy must be supplied to the rotor which is generated by the loss in potential energy as the aircraft descends.

If $V_D$ is the rate of descent and $v_i$ the induced velocity, then from actuator disc theory, the rotor thrust is given by

$$T = \rho A(V_D - v_i)2v_i \qquad (7.3)$$

where $\rho$ is the air density and $A$ is the rotor disc area.

The autorotative condition is that the torque at the rotor hub must be equal to zero, or some other predetermined value to allow for transmission losses and to supply power to the tail rotor, for manoeuvring. This condition is analysed using blade element theory and the expression for the thrust is obtained from a chordwise

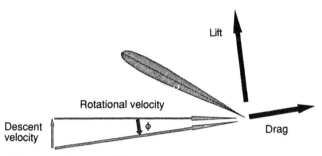

**Fig. 7.2.** Power-off descent flow condition.

strip of aerofoil. Equation (7.2) is modified by assuming $\phi$ is small giving $\cos\phi \approx 1$ and $\sin\phi \approx \phi$. Also assumed is a sensibly high value of lift/drag ratio for the blade aerofoil section.

With these assumptions, the thrust expression is given by

$$dT = dL \tag{7.4}$$

and the overall drag by

$$d\,\text{Drag} = dL \cdot \phi + dD \tag{7.5}$$

The contribution to the rotor thrust from $N$ aerofoil strips at radius $r$ and of width $dr$ is therefore

$$dT = \frac{1}{2}\rho Nc\,\Omega^2 R^2 x^2 Ra\left[\theta + \frac{V_D - v_i}{\Omega Rx}\right]dx \tag{7.6}$$

where $x$ is the non-dimensional rotor radius, $N$ is the number of blades, $c$ is the mean blade chord, $\Omega$ is the rotor speed, $R$ is the rotor radius and $a$ is the lift curve slope.

Integrating along the blade span over the range $0 \leqslant x \leqslant 1$ gives

$$T = \frac{1}{2}\rho Nc\,\Omega^2 R^3 a\left[\frac{\theta}{3} + \frac{1}{2}\frac{V_D - v_i}{\Omega R}\right] \tag{7.7}$$

Similarly the element of torque $dQ$ is given by

$$dQ = r \cdot d\,\text{Drag}$$
$$= \left(-\frac{1}{2}\rho Nc\,\Omega^2 R^4 x^3 a\left[\theta + \frac{V_D - v_i}{\Omega Rx}\right]\frac{V_D - v_i}{\Omega Rx} + \frac{1}{2}\rho Nc\,\Omega^2 R^4 x^3 C_{D0}\right)dx \tag{7.8}$$

where $C_{D0}$ is the blade aerofoil profile drag coefficient.

The first term of Eqn (7.8) is the forward component of the lift force, the second is the rearward profile drag.

Integrating (7.8) along the blade span as before gives

$$Q = -\frac{1}{2}\rho Nc\Omega^2 R^4 a\left[\frac{\theta}{3} + \frac{1}{2}\frac{V_D - v_i}{\Omega R}\right]\frac{V_D - v_i}{\Omega R} + \frac{1}{8}\rho Nc\Omega^2 R^4 C_{D0} \tag{7.9}$$

Expressing the thrust and torque in coefficient form we have

$$C_T = \frac{T}{\frac{1}{2}\rho(\Omega R)^2 \pi R^2}, \qquad C_Q = \frac{Q}{\frac{1}{2}\rho(\Omega R)^2 R\pi R^2} \tag{7.10}$$

from Eqns (7.7)–(7.10) we see

$$C_Q = -C_T\left[\frac{V_D - v_i}{\Omega R}\right] + \frac{1}{4}sC_{D0} \tag{7.11}$$

For a steady autorotative descent we must have

$$C_Q = 0 \tag{7.12}$$

From which (7.11) becomes

$$C_T\left[\frac{V_D - v_i}{\Omega R}\right] = \frac{1}{4}sC_{D0} \tag{7.13}$$

Equation (7.3) can be written in coefficient form

$$C_T = 2\left(\frac{V_D - v_i}{\Omega R}\right)2\frac{v_i}{\Omega R} \tag{7.14}$$

Eliminating the term

$$\frac{V_D - v_i}{\Omega R} \tag{7.15}$$

from (7.13) and (7.14) gives

$$\frac{C_T^2}{4}\frac{\Omega R}{v_i} = \frac{1}{4}sC_{D0} \tag{7.16}$$

From which

$$\frac{v_i}{\Omega R} = \frac{C_T^2}{sC_{D0}} \tag{7.17}$$

Substituting this result in (7.13) gives finally

$$\frac{V_D}{\Omega R} = \frac{1}{4}\frac{sC_{D0}}{C_T} + \frac{C_T^2}{sC_{D0}} \tag{7.18}$$

In steady vertical autorotation, assuming the pilot has operated his controls correctly, $T = W$ and $\Omega R$ is the normal operational value, hence $C_T$ is the same value as in hover and $V_D$ can be calculated. Low values of $s$ and $C_{D0}$ reduce $V_D$, since $C_T^2$ is small compared to $C_T$.

## Descent speed calculation

As an illustration of the calculation using eqn (7.18) two cases are shown below. They are for two early design cases in the development of the Westland WG13 helicopter, with a substantial difference in main rotor size, but other data common. The difference reflects different design priorities regarding land-based or ship-based operation. Case I is the utility Westland WG13 helicopter, initial design and case II is the naval Westland WG13 helicopter. Early proposals required a much reduced main rotor radius for adequate blade tip/hangar door clearance on board a ship. Table 7.1 gives details of the two cases. As can be seen, case II requires a much higher descent rate to achieve a steady autorotative condition.

**Table 7.1**

| Case | I | II |
|---|---|---|
| All-up weight (N) | 35669 | 35669 |
| No. of blades | 4 | 4 |
| Blade chord (m) | 0.394 | 0.394 |
| $C_{D0}$ | 0.011 | 0.011 |
| Main rotor radius (m) | 7.52 | 5.49 |
| Solidity | 0.0667 | 0.091 |
| $C_T$ | 0.0072 | 0.0135 |
| $\Omega R$ (m/s) | 213.4 | 213.4 |
| $V_D/\Omega R$ | 0.0255+0.0707 | 0.0185+0.182 |
| $V_D$ (m/s) | 20.5 | 42.8 |

## Transient behaviour in autorotation

The rotational speed of the rotor, corresponding to ideal autorotation and minimum descent rate, may be somewhat different to the rotor speed at the instant of power loss. The transient behaviour of the rotor speed, which is influenced by the rotor inertia and net torque, is important in producing a satisfactory entry into autorotation. If the pilot does not quickly remove the aerodynamic retarding torque by reducing collective pitch, then the rotor may slow down too much. A large rotor polar moment of inertia will help reduce the decay of the rotor speed. This will ensure in the steady autorotative descent, a maximum ratio of the kinetic energy of the rotor to the kinetic energy of the helicopter as a whole. The advantage of this is to build up an energy store in the rotor for the final part of the autorotative manoeuvre. This is the landing flare prior to touchdown, which is necessary to arrest the vertical descent rate of the helicopter. As Table 7.1 shows, in steady autorotation, considerable descent velocities occur which if sustained throughout the landing would cause immense damage to the airframe and flight crew. In preparing for a safe landing, the pilot raises the collective pitch in order to obtain the rotor thrust required to arrest the rapid descent rate of the helicopter. This puts the rotor back into powered flight state (as in Fig. 7.1) and since the power supply is still zero, the rotor speed will rapidly decrease. The pilot trades the kinetic energy of the rotor for the thrust necessary to give a sensible vertical speed, to ensure a safe touchdown.

As a final comment on the steady autorotative descent part of the manoeuvre, where the net torque must be zero, consider in Fig. 7.2 the forces on a blade element. The collective pitch angle $\theta_0$ must be such that there is a net forward component of force on some blade sections. However, when the rotational inflow speed is small, i.e. towards the blade root, the sections will be stalled and the profile drag consequently large, in which case the net force will be retarding. Near the tip, where the rotational inflow speed is large, the lift inclination will not be sufficiently forward and again a retarding force results. Hence it is the spanwise central regions of the blade which have the forward component of force necessary to drive the rotor.

## Landing flare

As has been described, it is necessary to arrest the rate of descent before the helicopter touches the ground and this is achieved by the pilot raising the collective pitch lever and generating rotor lift greater than the weight of the helicopter and thus decelerating it. The energy stored in the rotor is traded for lift. The condition of the rotor blade aerofoil section is now that of Fig. 7.1 rather than Fig. 7.2 and the rotor will therefore lose rotational speed. The only thing which can now keep the rotor turning is its kinetic energy and this has a limited lifetime as the energy is drained from it to obtain rotor thrust.

## Power failure near to the ground in near hovering flight

The requirement for a helicopter to make a safe power-off landing is that it has sufficient kinetic energy in the rotor. A simple example of this type of manoeuvre, or flare, is given below in the case of an engine power failure when the helicopter is hovering close to the ground.

If the helicopter is near to the ground when the power failure occurs, it is impossible for the pilot to establish a steady descent condition and to complete a conventional autorotative landing. The pilot can, in this case only, use the kinetic energy stored in the rotor at the instant of engine failure to reduce the rate of descent, and the rotor controls, particularly the collective pitch lever, are held in position. The safety of the landing will then depend on one or both of the following two factors: (a) the maximum descent velocity which the landing gear and fuselage can absorb and (b) the minimum permissible rotor speed which can be tolerated without the blade coning angle becoming excessive. As the flare is initiated the rotor is in a high thrust and low rotational speed condition. If the rotor speed is too low, then the blade flapping will become uncontrolled with the inevitable catastrophic consequences.

The height from which this type of descent can be safely made[15] is analysed as follows.

The equation of vertical motion is

$$\ddot{y} = g\left(1 - \frac{T}{W}\right) \tag{7.19}$$

where $T$ is the rotor thrust, $W$ is the helicopter weight, $y$ is the descent height from power failure ($t = 0$) and $g$ is the acceleration due to gravity.

The rotor rotation equation is

$$J\dot{\Omega} = Q_s - Q_a \tag{7.20}$$

where $J$ is the polar moment of inertia of the main rotor, $\Omega$ is the rotor speed, $Q_s$ is the shaft torque input from the engine(s) and $Q_a$ is the aerodynamic torque.

Immediately before the power failure we have

$$Q_s = Q_a, \qquad \dot{\Omega} = 0 \tag{7.21}$$

When the power fails

$$Q_s = 0, \qquad J\dot{\Omega} = -Q_a \tag{7.22}$$

It is a reasonable assumption that the rotor thrust, $T$, and the rotor torque, $Q$, vary with the square of the rotor speed, $\Omega$, hence

$$T = W\left(\frac{\Omega}{\Omega_0}\right)^2 \tag{7.23}$$

and

$$Q_a = Q_0\left(\frac{\Omega}{\Omega_0}\right)^2 \tag{7.24}$$

where $\Omega_0$ is the rotor speed at power failure and $Q_0$ is the steady power output immediately before failure.

Substitution of (7.24) in (7.22) gives

$$J\dot{\Omega} = -Q_0 \left(\frac{\Omega}{\Omega_0}\right)^2 \tag{7.25}$$

Now if we define

$$\overline{\Omega} = \frac{\Omega}{\Omega_0} \tag{7.26}$$

Equation (7.25) becomes

$$\dot{\Omega} = -\frac{Q_0}{J}\overline{\Omega}^2, \qquad \frac{d\overline{\Omega}}{dt} = -\frac{Q_0}{J\Omega_0}\overline{\Omega}^2 \tag{7.27}$$

and (7.19) becomes

$$\ddot{y} = g\left(1 - \overline{\Omega}^2\right) \tag{7.28}$$

Now if we define

$$\alpha = \frac{Q_0}{J\Omega_0} \tag{7.29}$$

Equation (7.27) reduces to

$$\frac{d\overline{\Omega}}{dt} = -\alpha\overline{\Omega}^2 \tag{7.30}$$

With the boundary condition of $\overline{\Omega} = 1$ at $t = 0$, the solution is

$$\overline{\Omega} = \frac{1}{1 + \alpha t} \tag{7.31}$$

Substituting this result into (7.28) gives

$$\ddot{y} = g\left(1 - \frac{1}{(1 + \alpha t)^2}\right) \tag{7.32}$$

which, with the boundary conditions $y(0) = \dot{y}(0) = 0$, gives

$$y = \frac{1}{2}gt^2 + \frac{g}{\alpha^2}\ln(1 + \alpha t) - \frac{g}{\alpha}t \tag{7.33}$$

To give a comparison of the effectiveness of this manoeuvre, consider an object, a brick say, falling under gravity. If $y_{\mathrm{BRICK}}$ is the height loss in the same time $t$ then we have

$$y_{\mathrm{BRICK}} = \tfrac{1}{2}gt^2 \qquad (7.34)$$

then expressing the height loss of the helicopter compared to the brick we have the ratio

$$\frac{y}{y_{\mathrm{BRICK}}} = 1 + \frac{2}{\alpha^2 t^2}\ln(1+\alpha t) - \frac{2}{\alpha t}$$

$$= \alpha t\left(\tfrac{2}{3} - \tfrac{1}{2}\alpha t + \ldots\right) \qquad (7.35)$$

$y/y_{\mathrm{BRICK}}$ is the distance which the helicopter falls compared to the distance which a body (brick) falling freely under $1g$ acceleration would have fallen in the same time.

It can be seen that from the $\alpha$ term outside the parentheses, the higher the polar moment of inertia of the rotor ($J$) and the normal operating rotor speed ($\Omega_0$) the greater will be the difference between $y$ and $y_{\mathrm{BRICK}}$ and the better will be its flare performance. An example using these results is presented in Figs 7.3–7.5. The rotor speed at the instant of power failure is 21.8 rad/s. The figures show the effect

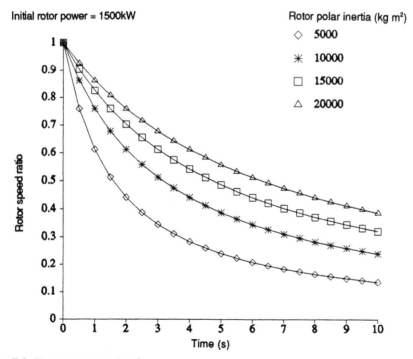

**Fig. 7.3.** Flare rotor speed ratio.

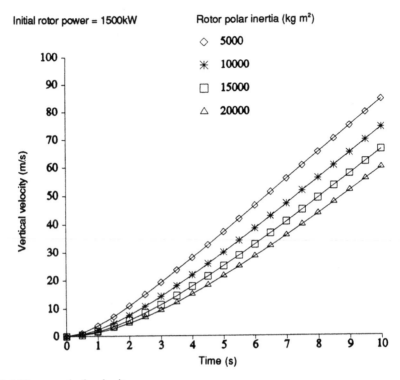

**Fig. 7.4.** Flare vertical velocity.

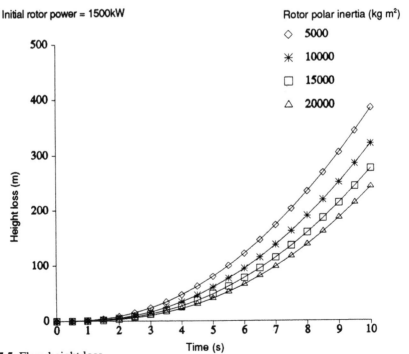

**Fig. 7.5.** Flare height loss.

of increasing polar inertia of the rotor. Figure 7.3 shows the rotor speed decay, Fig. 7.4 the vertical velocity increase and Fig. 7.5 the height loss. The basic conclusion is that this manoeuvre can only be contemplated for a limited time. If the polar inertia of 10 000 kg m$^2$ is selected, then in 2 s the rotor speed reduces to 61% of normal with a vertical velocity of 8 m/s, both of which are severe situations.

### *Power failure in forward flight*

When power failure occurs in forward flight, the pilot is able to lift the nose of the helicopter, using cyclic pitch, tilting the rotor up at the front, and so produce a flow of air up through the rotor. This supplies the energy to keep the rotor turning by using the kinetic energy associated with the forward speed. Tilting the rotor disc rearwards to achieve this causes a rearward tilt of the rotor thrust vector, and thereby increases the helicopter's drag. It is overcoming this drag which consumes the kinetic energy. This upflow of air through the rotor disc puts the blades into the condition of Fig. 7.2 rather than Fig. 7.1 so that the drop in rotor speed is much less than in the hovering condition. In fact, if the forward speed is at all moderate then no loss of rotor speed need occur.

The helicopter is then able to slow down, descend in a controlled manner and land using the kinetic energy in the rotor as previously described. Life is not always so simple since the helicopter attitude at the point of landing can be significantly nose up with the attendant problem of hitting the ground with the tail.

## Dead man's curve

For a helicopter suffering a total power loss there are flight conditions (height/ speed combination) from which it cannot descend and land safely. The conditions are usually described on a plot with forward speed as abscissa and altitude as ordinate. The problem areas are plotted on these axes in what is called the dead man's curve and an example is given in Fig. 7.6. The helicopter should not be operated inside the shaded zones of the height vs speed curve if total power failure is not to be allowed to hazard the aircraft since a safe power-off landing is not possible.

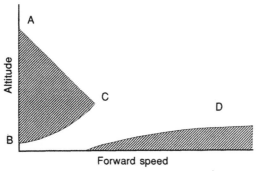

**Fig. 7.6.** Dead man's curve.

The various points on the curve are discussed below

A   This is the minimum height from which a hovering helicopter can establish a steady descent velocity and build up its rotor rotational speed sufficiently to make a safe landing.

B   This is the maximum height from which a helicopter can land using its rotor kinetic energy to restrict the rate of descent at touchdown and/or its minimum rotor speed at touchdown.

C   At this speed and height, the pilot can use the kinetic energy of motion to maintain rotor speed and make a safe landing.

D   This region is a height/speed band where the speed of the helicopter is too high to make a run on landing and where a sharp nose up flare will cause the aircraft to balloon up in altitude or cause the tail rotor to strike the ground, because of the high nose up attitude of the fuselage at low altitude.

## Multi-engined flight following power failure of one engine

The performance of a helicopter following partial power failure depends upon the level of the remaining power in the helicopter.

Figure 7.7 shows the normal power vs forward speed variation of a helicopter. If the power remaining after one engine has failed is represented by the line $YY$ then the helicopter is able to continue with any manoeuvre provided that a further engine failure does not occur. If, however, the situation following failure of one engine is represented by the line $XX$ then the helicopter will only be able to maintain level flight between the speeds $V_1$ and $V_2$. Below $V_1$, the helicopter will be able to use the engine torque available to reduce the height $A$ in Fig. 7.6, and to increase the height of point $B$. The dead man's curve will decrease in area until $V_1 = 0$, when it will disappear altogether.

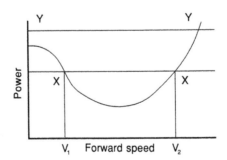

**Fig. 7.7.** Effect of engine failure on a twin-engined helicopter.

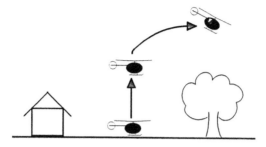

**Fig. 7.8.** Towering take-off.

# Take-off techniques allowing for engine failure

In order to obtain an airworthiness certificate, a helicopter must be able to operate safely. This must include a power failure and still comply with the dead man's curve which has been described, and which defines areas of operation which should be avoided. However, take off techniques can be used to allow the helicopter to take off and proceed to cruising flight and always remain in the safe areas of dead man's curve.

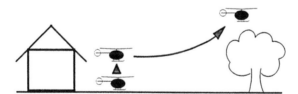

**Fig. 7.9.** Forward take-off.

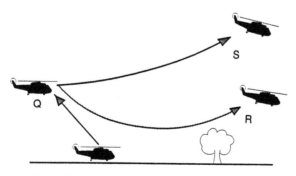

**Fig. 7.10.** Backward take-off.

Three take-off techniques (a) (b) (c) are shown in Figs 7.8–7.10, respectively

(a)  The towering or near vertical take-off is unsafe unless the helicopter has sufficient power with one engine failed to meet condition *YY* of Fig. 7.7.

(b)  The forward take-off is safe provided that the flight path is arranged so that height and speed is always below *BC* and above *D* of Fig. 7.6. This may still not be completely safe because of obstructions ahead which prevent the helicopter landing following total power failure. The pilot quickly loses view of the take-off point with this technique. Use of ground effect, by accelerating the helicopter close to the ground, can help this technique.

(c)  The backward take-off is used in civil multi-engined helicopters. Point *Q* is the decision height. Below *Q* the helicopter lands back on the site with a low forward speed following one engine failure. At *Q*, if an engine fails, the helicopter is able to accelerate to the speed $V_1$ of Fig. 7.7 and then climb away along the line *QR* without hitting any obstructions. Normal take-off path with full power available is *QS*.

# 8

# Blade dynamics

## Dynamics of a beam

In the discussion on blade dynamics so far, the blade has been considered to be rigid. In reality, the blade will bend elastically and its response to the various aerodynamic forces will need to be calculated. In this chapter the blade is modelled by means of beam theory and the concept of modal methods described. The rigid blade approximation is usually sufficient for an overall treatment of rotor disc control, however, as the frequency of the aerodynamic variation increases so the elastic behaviour of the rotor blade becomes increasingly important and therefore must be introduced into the theoretical methods.

The three degrees of freedom of the blade, namely flap, lag and torsion, are usually treated together in what is termed coupled mode methods. For this introduction to the subject only flapwise deflections and forcings are considered. To analyse a moving helicopter blade, the dynamic behaviour of a flexible beam is derived. The forces and moments acting on a typical blade element $AB$ are shown in Fig. 8.1. (The motion is undamped.)

## Stationary beam ($\Omega = 0$)

Initially the dynamic equations of motion are derived for a stationary blade. Resolving vertically

$$S + dS + w \cdot dr - S = 0 \tag{8.1}$$

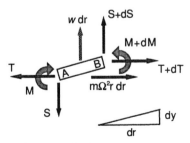

**Fig. 8.1.** Forces and moments acting on a blade element. $S$ is the shear force, $T$ the tension, $M$ the bending moment, $w$ the loading (assumed to act the centre of the element) and $m$ is the mass distribution.

which gives on proceeding to the limit as $\delta r \to 0$

$$\frac{\mathrm{d}S}{\mathrm{d}r} = -w \tag{8.2}$$

Resolving horizontally

$$T - (T + \mathrm{d}T) = 0 \tag{8.3}$$

which results in

$$T = 0 \tag{8.4}$$

Taking anticlockwise moments about $A$

$$M + \mathrm{d}M + (S + \mathrm{d}S) \cdot \mathrm{d}r + (w\,\mathrm{d}r) \cdot \frac{\mathrm{d}r}{2} - M = 0 \tag{8.5}$$

from which

$$\frac{\mathrm{d}M}{\mathrm{d}r} = -S \tag{8.6}$$

and from (8.2)

$$\frac{\mathrm{d}^2 M}{\mathrm{d}r^2} = w \tag{8.7}$$

Recalling the beam equation

$$M = EIy'' \tag{8.8}$$

where the prime represents differentiation with respect to $r$. We find

$$[EIy'']'' = w \tag{8.9}$$

In fact, $w$ consists of the upward external forcing to the element, $W$, and the downward inertial forcing, $m\ddot{y}$, i.e.

$$w = W - m\ddot{y} \tag{8.10}$$

from which the final equation of motion becomes

$$[EIy'']'' = W - m\ddot{y} \tag{8.11}$$

The external forcing $W$ and the response $y$ are, strictly, functions of span and time. To analyse the equation of motion the forcing is expressed as a product of a forcing shape $F(x)$, applied sinusoidally at a circular frequency $p$. The response is similarly expressed as a product of a deflection shape $Y(r)$, also at a frequency of $p$. The form of the equation of motion justifies this assumption of identical frequency.

Mathematically we have

$$W(r,\ t) = F(r) \cdot \cos pt$$
$$y(r,\ t) = Y(r) \cdot \cos pt \tag{8.12}$$

Substituting these results into (8.11) and cancelling the $\cos pt$ term we find

$$[EI \cdot Y'']'' - p^2 mY = F \tag{8.13}$$

# Introduction of modes

Any beam which is not fixed will oscillate in a series of natural modes which can be designated by $y_k$ for the $k$th mode which has frequency $p_k$ and this satisfies the equation

$$[EIy_k'']'' - p_k^2 my_k = 0 \tag{8.14}$$

The solution to be sought is a linear combination of the mode shapes, the multiplying coefficients being the responses to the individual modes. Namely

$$Y(r) = \sum_1^\infty C_k \cdot y_k(r) \tag{8.15}$$

This expression is substituted into Eqn (8.13) which gives

$$\sum_k C_k [EI \cdot y_k'']'' - p^2 m \sum_k C_k y_k = F \tag{8.16}$$

The problem arises of an infinite number of unknowns, i.e. $C_k$, $k = 1,\ 2,\ 3,\ \dots$ . A means has to be devised to isolate the individual terms, $C_k$, and enable each individual modal response to be determined.

The algebraic equation (8.16) can be converted into an infinite series of simultaneous equations by multiplying it by each mode, $y_n$, successively and integrating with respect to $r$ along the blade length, i.e.

$$\int_0^R [\text{eqn (8.16)}] \cdot y_n\, \mathrm{d}r \tag{8.17}$$

This removes the $r$ dependence, but the problem of an infinite set of unknowns remains. This impasse can be broken if each of the equations, so produced, simplifies to one term on the left-hand side only. In this way the response to one mode can be determined. Each response can then be calculated by applying this scheme for each mode $y_n$ in turn. The isolation of each modal response requires that all terms in other modes vanish from the left-hand side of the equation. Such a property is known as orthogonality, and modes which satisfy Eqn (8.14) do indeed possess this property. The algebraic statement of orthogonality is

$$\int_0^R my_i \cdot y_j \, dr = 0, \quad i \neq j \tag{8.18}$$

Note that the mass distribution term $m$ is explicitly included in the integrand, the necessity of which will be shown. This is described as orthogonality of the modes with respect to the mass distribution. This can be proved by returning to Eqn (8.14), for two indices namely $k$ and $n$. A lemma is needed, i.e.

$$\int_0^R [EIy_k'']'' y_n \, dr = \int_0^R EIy_k'' y_n'' \, dr \tag{8.19}$$

Integrating the left-hand side by parts gives

$$\int_0^R [EIy_k'']'' y_n \, dr = \left| [EIy_k'']' y_n \right|_0^R - \int_0^R [EIy_k'']' y_n' \, dr \tag{8.20}$$

The first term on the right-hand side vanishes at the root ($r = 0$) because of zero displacement, $y_n$, and the tip ($r = R$) because of zero shear force $[EIy_k'']'$. Integrating by parts again gives

$$-\int_0^R [EIy_k'']' y_n' \, dr = -\left| [EIy_k''] y_n' \right|_0^R + \int_0^R EIy_k'' y_n'' \, dr \tag{8.21}$$

The first term on the right-hand side also vanishes at the root because of either zero slope, $y_n'$, or bending moment, $EIy_k''$, and at the tip because of zero bending moment, $EIy_k''$. Thus a pinned–free (*articulated blade*), or a built-in–free (*semi-rigid blade*) beam satisfies these boundary conditions and the lemma is proved.

Now consider the equations for two modes of index $k$ and $n$

$$[EIy_k'']'' - p_k^2 my_k = 0$$
$$[EIy_n'']'' - p_n^2 my_n = 0 \tag{8.22}$$

Multiplying each modal equation by its alternate mode shape and integrating along the blade gives

$$\int_0^R [EIy_k'']'' y_n \, dr - p_k^2 \int_0^R my_k y_n \, dr = 0$$

$$\int_0^R [EIy_n'']'' y_k \, dr - p_n^2 \int_0^R my_n y_k \, dr = 0$$

(8.23)

Noting that from the lemma

$$\int_0^R [EIy_k'']'' y_n \, dr = \int_0^R EIy_k'' y_n'' dr$$

$$= \int_0^R EIy_n'' y_k'' dr$$

(8.24)

$$= \int_0^R [EIy_n'']'' y_k \, dr$$

means that the second term of the left-hand sides of (8.23) must be equal, which re-expressed becomes

$$(p_k^2 - p_n^2) \int_0^R my_k y_n \, dr = 0$$

(8.25)

If the modal frequencies $p_k$ and $p_n$ have different values, then $p_k^2 - p_n^2 \neq 0$, and can be cancelled, from which the orthogonality condition is proved.

The above analysis also gives the following result from (8.23) and (8.24)

$$\int_0^R EIy_k'' y_n'' dr = 0, \quad k \neq n$$

(8.26)

The modal equation also allows an expression to be derived which explicitly links a mode shape and its frequency together.

Returning to (8.14) we obtain the following by multiplying the $n$th modal equation by its own mode shape, $y_n$, and integrating along the blade

$$\int_0^R [EIy_n'']'' y_n \, dr - p_n^2 \int_0^R my_n^2 \, dr = 0$$

(8.27)

Using Eqn (8.24) gives the following equation which links the mode shape and frequency explicitly together

$$p_n^2 = \frac{\int\limits_0^R EIy_n''^2 \, dr}{\int\limits_0^R my_n^2 \, dr} \tag{8.28}$$

This is known as Rayleigh's formula.

Returning to the solution of the modal responses, the orthogonality is proved and can now be used in the solution. Recalling (8.16) and (8.17) the following result is obtained

$$\int\limits_0^R \sum_k C_k [EIy_k'']'' \cdot y_n \, dr - p^2 \int\limits_0^R \sum_k C_k my_k y_n \, dr = \int\limits_0^R F \cdot y_n \, dr \tag{8.29}$$

Exchanging integration and summation gives

$$\sum_k C_k \int\limits_0^R [EIy_k'']'' y_n \, dr - p^2 \sum_k C_k \int\limits_0^R my_k y_n \, dr = \int\limits_0^R F \cdot y_n \, dr \tag{8.30}$$

Because of (8.18) and (8.26) only the $n$th terms of the summations in (8.30) remain, i.e.

$$C_n \int\limits_0^R [EIy_n'']'' y_n \, dr - p^2 C_n \int\limits_0^R my_n y_n \, dr = \int\limits_0^R F \cdot y_n \, dr \tag{8.31}$$

Use of previous results [Eqns (8.24) and (8.28)] finally gives

$$C_n = \frac{1}{p_n^2 - p^2} \frac{\int\limits_0^R Fy_n \, dr}{\int\limits_0^R my_n^2 \, dr} \tag{8.32}$$

The final solution can now be assembled giving

$$y(r, t) = \sum_k \frac{1}{p_k^2 - p^2} \frac{\int\limits_0^R Fy_k \, dr}{\int\limits_0^R my_k^2 \, dr} \cdot y_k(r) \cdot \cos pt \tag{8.33}$$

This result shows three distinct methods of reducing a particular mode's response to a given forcing

(a) Keep the natural frequencies of the modes away from the forcing frequency. This is essentially an avoidance of resonance. Maximise $p_k^2 - p^2$.
(b) Design the mode shapes to be orthogonal to the forcing shape. Minimise $\int F y_k \, dr$.
(c) Concentrate the blade mass, in a spanwise sense, at antinodes. This is equivalent to maximising the modal inertia which is defined by

$$\text{modal inertia,} \quad I_n \triangleq \int_0^R m y_n^2 \, dr \qquad (8.34)$$

## Rotating beam ($\Omega \neq 0$)

The stationary beam analysis can be extended to cater for a rotating beam. The difference is in the variable tension loads in the beam (refer to Fig. 8.1). Resolving vertically

$$S + dS + w \, dr - S = 0 \qquad (8.35)$$

Giving as $\delta r \to 0$

$$\frac{dS}{dr} = -w \qquad (8.36)$$

Resolving horizontally

$$T + dT + m\Omega^2 r \, dr - T = 0 \qquad (8.37)$$

Giving

$$\frac{dT}{dr} = -m\Omega^2 r \qquad (8.38)$$

which can be integrated to

$$T = \Omega^2 \int_r^R m\eta \, d\eta \qquad (8.39)$$

Taking anticlockwise moments about $A$

$$M + dM + (S + dS) \cdot dr + w \, dr \cdot \frac{dr}{2} - (T + dT) \cdot dy - M = 0 \qquad (8.40)$$

we have

$$\frac{\mathrm{d}M}{\mathrm{d}r} + S - T\frac{\mathrm{d}y}{\mathrm{d}r} = 0 \tag{8.41}$$

Assembling the final equation and again noting that the forcing $w$ consists of an external forcing $W$ and an inertial term $m\ddot{y}$

$$\frac{\mathrm{d}^2 M}{\mathrm{d}r^2} = [EIy'']''$$

$$= W - m\ddot{y} + \frac{\mathrm{d}}{\mathrm{d}r}\left(T\frac{\mathrm{d}y}{\mathrm{d}r}\right) \tag{8.42}$$

$$= W - m\ddot{y} + (Ty')'$$

i.e.

$$[EIy'']'' - (Ty')' = W - m\ddot{y} \tag{8.43}$$

hence with a forcing $F$ of circular frequency $p$, as before, the response shape, $Y$, satisfies the equation

$$[EIY'']'' - (TY')' - p^2 mY = F \tag{8.44}$$

The equation for a rotating mode now becomes

$$[EIy_k'']'' - (Ty_k')' - p_k^2 my_k = 0 \tag{8.45}$$

The modes can be shown to be orthogonal in an identical manner to that already described for the non-rotating modes. The extra term involving the blade axial tension will need to be addressed, and the following expression will occur

$$\int_0^R (Ty_k')'y_n \,\mathrm{d}r \tag{8.46}$$

Integrating by parts gives

$$\int_0^R (Ty_k')'y_n \,\mathrm{d}r = \left|Ty_k'y_n\right|_0^R - \int_0^R Ty_k'y_n' \,\mathrm{d}r \tag{8.47}$$

The first term on the right-hand side vanishes at both limits because of either zero deflection, $y_n$, at the root or zero blade tension, $T$, at the tip. This gives the following result

$$\int_0^R (Ty_k')'y_n\, dr = -\int_0^R Ty_k'y_n'\, dr$$

$$= \int_0^R (Ty_n')'y_k\, dr \tag{8.48}$$

Using this result the following orthogonality relations can be proved

$$\int_0^R my_k y_n\, dr = 0, \quad n \neq k$$

$$\int_0^R (EIy_k'')''y_n\, dr - \int_0^R (Ty_k')'y_n\, dr = 0, \quad n \neq k \tag{8.49}$$

If the $n$th modal equation (8.45) is multiplied by its own mode shape, $y_n$, and then integrated along the blade, the following result is obtained. In the non-rotating case this led to Rayleigh's formula which gives the modal frequency directly in terms of the mode shape and the blade structural characteristics

$$\int_0^R [EIy_n'']''y_n\, dr - \int_0^R (Ty_n')'y_n\, dr - p_n^2 \int_0^R my_n^2\, dr = 0 \tag{8.50}$$

From which, using (8.24) and (8.48) we have

$$\int_0^R EIy_n''^2\, dr + \int_0^R Ty_n'^2\, dr - p_n^2 \int_0^R my_n^2\, dr = 0 \tag{8.51}$$

which finally gives

$$p_n^2 = \frac{\int_0^R EIy_n''^2\, dr}{\int_0^R my_n^2\, dr} + \frac{\int_0^R Ty_n'^2\, dr}{\int_0^R my_n^2\, dr} \tag{8.52}$$

Recalling Eqn (8.28), the Rayleigh formula, it can be seen that the first term of (8.52) is identical to the equivalent frequency in the non-rotating case. If this is denoted by $p_{n0}$, then

$$p_{n0}^2 = \frac{\int\limits_0^R EIy_n''^2 \, dr}{\int\limits_0^R my_n^2 \, dr} \tag{8.53}$$

Also recalling Eqn (8.39), which gives the blade tension in terms of rotor speed and blade mass distribution, (8.52) can be re-expressed as

$$p_n^2 = p_{n0}^2 + \Omega^2 \left[ \frac{\int\limits_0^R y_n'^2 \left( \int\limits_r^R m\eta \, d\eta \right) dr}{\int\limits_0^R my_n^2 \, dr} \right] \tag{8.54}$$

The frequency of the $n$th mode can therefore be expressed as a combination of a component due to the elastic properties of the blade, $p_{n0}$, and another which depends on the blade tension effects which is the centrifugal stiffening. In earlier chapters, the flapping motion of a semi-rigid rotor blade was analysed using an equivalent flapping hinge and a spring to adjust the frequency. The expression is of the form

$$\omega_\beta^2 = \omega_{\beta 0}^2 + \Omega^2 \left( 1 + \frac{3e}{2} \right) \tag{8.55}$$

which is of identical form to Eqn (8.54). If the more detailed analysis is used with a rigid blade mode then the two expressions (8.54) and (8.55) will converge to the same expression, linking the two analyses.

In general, the natural frequency of a mode is of the form

$$p_n^2 = p_{n0}^2 + K\Omega^2 \tag{8.56}$$

The analysis of the forced blade motion can now be solved in an identical fashion to the non-rotating case already discussed. The orthogonality conditions are proved and the responses to the modes, i.e. the $C_n$ terms will decouple, and the final solution follows as before. If we again seek a solution of the form

$$y = \sum_k C_k y_k \tag{8.57}$$

then substituting in (8.44) gives

$$\sum_k C_k [EIy_k'']'' - \sum_k C_k(Ty_k')' - p^2 m \sum_k C_k y_k = F \qquad (8.58)$$

multiplying by the $n$th mode shape, $y_n$, and integrating along the blade gives

$$\sum_k C_k \int_0^R [EIy_k'']'' y_n \, dr - \sum_k C_k \int_0^R (Ty_k')' y_n \, dr - p^2 \sum_k C_k \int_0^R m y_k y_n \, dr = \int_0^R F y_n \, dr \quad (8.59)$$

Hence

$$\sum_k C_k \left[ \int_0^R [EIy_k'']'' y_n \, dr - \int_0^R (Ty_k')' y_n \, dr \right] - p^2 \sum_k C_k \int_0^R m y_k y_n \, dr = \int_0^R F y_n \, dr \quad (8.60)$$

Using the previous results, we find that only the terms where $k = n$ remain, hence

$$C_n \left[ \int_0^R EIy_n''^2 \, dr + \int_0^R Ty_n'^2 \, dr - p^2 \int_0^R m y_n^2 \, dr \right] = \int_0^R F y_n \, dr \qquad (8.61)$$

Eliminating the first two terms of (8.61) using (8.51) we have

$$C_n = \frac{1}{p_n^2 - p^2} \frac{\displaystyle\int_0^R F y_n \, dr}{\displaystyle\int_0^R m y_n^2 \, dr} \qquad (8.62)$$

Provided that the modal frequencies and shapes are known, the solution of the rotating beam response is no more complicated than if the beam were stationary. The extra difficulty, of course, is the added complication of the blade tension term in the modal equation.

## Spoke, interference or Campbell diagram

The variation of the modal frequency with rotor speed will cause some difficulties for the helicopter rotor dynamicist in terms of avoiding possible resonant conditions. Here the blade natural frequency will be close to, or indeed equal to, that of the aerodynamic forcing effects. The spoke or Campbell diagram allows any possible frequency coalescences to be predicted. It assumes that typical aerodynamic forcing frequencies are multiples of the rotor speed $\Omega$. The diagram is a plot of frequency, as ordinate, against rotor speed, as abscissa. The forcing frequencies ($n\Omega$) appear as straight lines and the modal frequency variation described in Eqn

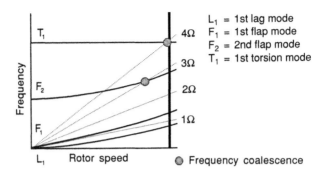

**Fig. 8.2.** Spoke, interference or Campbell diagram.

(8.54) can also be plotted (see Fig. 8.2). Crossings of the two frequency plots indicate possible resonant conditions. Clearly these must be avoided near to the operating rotor speed. The stability of the blade motion at coincident frequencies below the operating rotor speed has to be investigated for particular rotor systems. In difficult cases, operation will require that the rotors need to be accelerated or decelerated through these frequency values.

## Lag and torsion motions

The preceding analysis is concerned with flapwise deflections, however, similar methods can be applied to lag and torsion motions. The torsion natural frequencies are normally high compared to the lower modes of flap and lag. Torsion motion differs from flap and lag in that the stiffness of the control system has a large influence.

In the lag plane, the centrifugal stiffening plays a similar effect to that in the flapping plane, however, as already discussed in Chapter 6, there is a reduction in the moment arm of the restoring moment in lag compared to flap and, consequently, the fundamental lag frequency will be lower than the fundamental in flap. Conversely, as the mode number increases the blade elastic bending becomes of increasing importance. As lag motion is accompanied by edgewise bending and flap by flapwise bending then the higher harmonics in lag normally possess higher frequencies than the corresponding flap modes.

## Finite element methods

As a postscript to this chapter, mention is made of finite element type of methods for the dynamic analysis of a structure. This could be a rotor blade or a complete fuselage. The methods so far described are of the "closed-form" type which are undoubtedly elegant but lack flexibility. The finite element method divides the structure into elements over which various properties can be assumed to be uniformly distributed, or concentrated at a point. The many interconnections between the elements allow a set of differential equations to be set up which can then be solved to obtain the mode shapes and frequencies. Each element will have translational and, possibly, rotational degrees of freedom and the total number of these summed over all the elements can become extremely large. Modern computers have the ability to solve such a massive set of equations and have allowed structures to be analysed in considerable depth and models possessing 40 000 degrees of freedom are now used.

Finite element methods are a considerable subject in their own right and a detailed description is beyond the scope of this book. However, they are mentioned to encourage the reader to investigate further.[16]

# Helicopter trim

## Control requirements

The ability to control completely the position and attitude of the helicopter in space requires control of the forces and moments about all three axes. The main rotor provides the longitudinal and lateral propulsive forces which means that there will be coupling between longitudinal forces and pitching moments, and between lateral forces and rolling moments. This removes independence from two of the six degrees of freedom. In other words, the ability to maintain longitudinal force equilibrium and to rotate in pitch to attain a desired attitude at will is sacrificed. Therefore, the pilot has four independent controls:

- *Vertical*—by changing the main rotor blade pitch, giving direct lift control through the main rotor thrust.
- *Yaw*—by changing the tail rotor blade pitch and thereby varying the tail rotor thrust.
- *Lateral*—by changing the direction of the main rotor thrust vector producing both a side force and a rolling moment.
- *Longitudinal*—by changing the direction of the main rotor thrust vector producing both a longitudinal force and a pitching moment.

Cross-coupling forces and moments are undesirable, but are usually avoidable. For instance, an increase in main rotor thrust necessitates an increase in rotor torque which requires a correction to the directional control. Such cross-referencing of activity calls for considerable coordination by the pilot. The work load of the pilot can be eased by a certain amount of mixing of the control inputs. The above example of a change in main rotor thrust and hence torque requiring a tail rotor thrust change to maintain heading can be accomplished by a mechanical coupling between the main rotor collective pitch lever and the pitch change mechanism to the tail rotor.

## Pilot's controls

The pilot is provided with the following controls:

- Cyclic stick (pilot's right hand). This provides longitudinal and lateral control. The stick is pushed in the required direction of flight, i.e. forward for a forward tilt of the main rotor disc. Plate 9.1 shows the layout of a helicopter cockpit with the pilot seated. The thumb is on a small control, called a beeper,

**Plate 9.1.** The pilot seated at the controls of a Westland Sea King helicopter. (© British Crown Copyright 1993/MOD reproduced with the permission of the Controller of Her Britannic Majesty's Stationery Office.)

which enables the pilot to adjust the aircraft trim attitude. This avoids the necessity of the pilot holding the cyclic stick in an off-centre position against a centring spring with the consequent improvement in comfort over a period of time.

- Pedals (pilot's feet). These provide yaw or directional control. The pedal is pushed in the required direction, i.e. left pedal for a port turn.
- Collective lever (pilot's left hand). This is the main rotor blade pitch control for vertical flight. The lever is operated in the required direction, i.e. upward movement for a thrust increase/aircraft climb.
- Engine throttle (pilot's left hand). If an independent throttle is provided, then it is usually situated on the collective lever as a twist grip. Normally, independent linkages produce the necessary throttle changes with adjustments in collective pitch so that the rotor speed is kept essentially constant. To alleviate the pilot of this control input, a constant speed governor can be fitted to match the power supplied by the engine to the new blade pitch setting. This option is more usual, and during operation the governor senses a change in the rotor speed away from the datum value and adjusts the power output from the engine(s) accordingly. (This topic is discussed in detail in Chapter 13.)

## Rotor behaviour in forward flight

Of paramount importance in the calculation of helicopter trim is the behaviour of the main rotor and fuselage in forward flight. The main rotor demonstrates two types of behaviour, namely

1. Effect of forward speed—this has already been discussed and is the rearward disc tilt induced by the dissymmetry of lift between the advancing and retreating sides of the disc. The blade flapping behaviour becomes upward at the front of the disc, and downward at the rear. This disc tilt, if uncorrected, increases with forward speed and the accompanying rearward tilt of the main rotor thrust vector gives a deceleration force and a nose-up pitching moment to the fuselage.
2. Effect of incidence—if the pitch attitude of the main rotor is disturbed then a component of the forward speed will be created normal to the rotor plane. This will induce an incidence change for the blades and hence the individual blade lift forces. For example, if the rotor tilts rearwards and the normal velocity through the rotor disc is then perturbed by an amount $\Delta U_P$, an incidence change of $\Delta \alpha$ will be induced where

$$\Delta \alpha = \frac{\Delta U_P}{U_T} \tag{9.1}$$

where $U_P$ and $U_T$ are the velocity components normal and tangential to the blade span, as used in previous chapters.

The blade lift will then alter by an amount proportional to

$$U_T^2 \cdot \Delta\alpha = \Delta U_P \cdot U_T \qquad\qquad (9.2)$$

the lift change will thus depend on the tangential velocity $U_T$ which means that the advancing blade will experience a greater lift increment than the retreating blade. The change in lift will give a change in flapping behaviour and because of the previous discussion the blades on the advancing side will see an upward flapping forcing with the opposite effect on the retreating side. The net result is an increase in coning but an additional rearward tilt of the rotor disc. This will cause a nose-up moment on the fuselage and as the fuselage reacts to this, the rotor disc will follow the shaft and tilt still further rearward. This will, in turn, cause an increase in the upflow through the rotor disc and the attendant *further* increase in rearward disc tilt. In a similar manner a forward tilt of the rotor disc will drive the fuselage nose down which, in turn, will cause the rotor disc to tilt even further forward. This is plainly unstable and can be passively controlled by means of creating a sufficiently large and correcting weathercock stability to the fuselage by means of a tailplane.

## Rotor control

As shown in Fig. 6.22, as has previously been discussed, there exists a plane in which the blade pitch angle remains constant called the *no feathering plane*. If cyclic pitch is applied then this plane will tilt by the same amount as the cyclic pitch and in consequence the plane is sometimes referred to as the control plane. It must not be confused with the plane of the swash plate which normally tilts predominantly laterally for the application of cyclic pitch to tilt the disc forwards. Two other planes can be used to define the rotor blade behaviour, namely that perpendicular to the rotor shaft (shaft axis) and that defined by the plane swept out by the rotor blade tips (tip path plane).

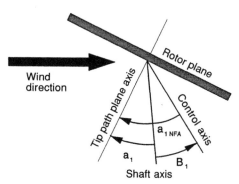

**Fig. 9.1.** Geometry of reference planes.

In summary

- *Tip path plane axes.* Since the resultant rotor force is virtually perpendicular to the tip path plane, this plane or axis system is very useful for quantifying the thrust and induced velocity, since no flapping occurs in this plane.
- *Control axes.* The control axes lie in the control plane and no feathering occurs about this plane. In reality the linkage system between the rotor blade pitch horn and the swash plate can mean that the physical position of the swash plate does not lie in the control plane.
- *Shaft axes.* Both flapping and feathering motions occur about this axis, but the shaft does provide an important physical datum for both the flapping and feathering control angles.

The control axis tilt is effected by the cyclic pitch control and as has been already explained, the control becomes more sensitive as forward speed increases.

In a torsionally rigid blade the feathering motion is applied by the control linkage and the root pitch angle is given by

$$\theta = \theta_0 - A_1 \cos\psi - B_1 \sin\psi \qquad (9.3)$$

where $\theta_0$ is the collective pitch angle and is dependent on the required rotor performance. $A_1$ is the lateral cyclic pitch angle and $B_1$ is the longitudinal cyclic pitch angle, both depending on the helicopter trim.

Recalling Chapter 5 the flapping angle is given by

$$\beta = a_0 - a_1 \cos\psi - b_1 \sin\psi \qquad (9.4)$$

where $a_1$ is the *rearward* disc tilt, and $b_1$ is the disc tilt *advancing side down*. Figure 9.1 shows the relative position of the various control axes.

The transformation between the no feathering axis (NFA) and the shaft axis is geometrical and is presented below

$$\theta_{0\,NFA} = \theta_0$$
$$a_{1\,NFA} = a_1 + B_1 \qquad (9.5)$$
$$b_{1\,NFA} = b_1 - A_1$$

The values of the three control angles, $\theta_0$, $A_1$ and $B_1$ for a given trim condition must now be determined.

## Helicopter trim

For a given all-up weight (AUW) and parasite drag at a given speed, the resultant main rotor thrust vector is fixed in space. This means that the disc tilt or cyclic flapping is completely determined for a given condition of operation (i.e. thrust,

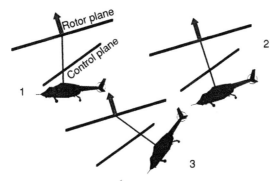

**Fig. 9.2.** Possible inclination of the rotor shaft.

forward speed, parasite drag or tip speed). The amount of cyclic pitch applied will determine how the disc is inclined relative to the control axis. Since the cyclic pitch defines the orientation of the control plane relative to the rotor shaft, the control axis inclination, relative to the flight path, can be achieved by various combinations of shaft angle and cyclic pitch control, as shown in Fig. 9.2. In practise, there are geometric limitations such as flapping stops. In addition, the fuselage aerodynamic characteristics must also be considered.

The figure shows three flight conditions varying from

1.  A level fuselage with a considerable control axis tilt, relative to it, and high values of cyclic pitch to give the required disc tilt.
2.  A more realistic situation with a nose down fuselage attitude and a moderate control axis tilt relative to it.
3.  A large nose-down fuselage attitude with a rearward tilt of the control axis relative to the fuselage. This can occur when the fuselage develops a large nose-down pitching moment and the rearward disc attitude relative to the fuselage is required to provide the correcting nose-up pitching moment.

In Fig. 9.2 for all three cases the inclination of the disc and control planes are identical, only the fuselage attitude varies. With no input from a hub moment or fuselage effect, any of these cases is a possible trim position. The introduction of hub and fuselage effects will mean that one fuselage attitude, control plane and disc plane combination will satisfy the trim conditions of no net forces or moments. An example of such a calculation is presented at the end of this chapter.

## Control power of a rotor

Controlling the feathering motion causes cyclic flapping (or equivalently a disc tilt) with its consequent thrust vector tilt relative to the shaft. (The resultant aerodynamic force of the rotor is approximately perpendicular to the disc.) The control of the whole helicopter is achieved by inclining the thrust vector and producing a moment about the helicopter CG. In the case of a hingeless rotor, this thrust moment is combined with a strong hub moment.

## Trim calculation

The control moment to be applied by the rotor depends upon the equilibrium conditions of the aircraft. The balance of forces and moments determine the position of the disc with respect to the helicopter. The attitude of the helicopter and the inclination of the disc relative to the helicopter flight direction (wind axis) is determined by the moment equilibrium about the CG. This is influenced by the amount of flapping which is controlled by the cyclic pitch. The independent controls are the collective pitch, which mainly affects the thrust and power demand of the rotor and the cyclic pitch, which mainly affects the rotor and helicopter attitude. As the main rotor thrust is used to control the helicopter in both the longitudinal and lateral planes there is a strong coupling via the rotor thrust between the longitudinal and lateral forces and moments so that no exact closed-form trim solution is possible. However, much useful work can be done by uncoupling the equations to give an approximate solution. To show this, the longitudinal trim is now considered. Although not included here, the lateral trim can be analysed in a similar manner, however, in the lateral plane there is the tail rotor thrust which is an important difference.

### *Longitudinal trim*

*Notation.* With reference to Fig. 9.3, the rotor is located at point $R$, the aircraft CG at point $G$, and the fuselage aerodynamic forces and moments based on an origin at point $A$, $T$ is the main rotor thrust, $M_s$ is the main rotor hub moment per unit disc tilt $(a_1)$, $L_F$ is the fuselage lift force (in wind tunnel axes), $D_F$ is the fuselage drag force (in wind tunnel axes), $M_F$ is the fuselage pitching moment and $W$ is the helicopter all-up weight. The sign convention for locations is positive upwards for waterlines, aft for stations and nose-up for pitch angles. For this analysis the rotor drag (or $H$ force) is neglected purely to simplify this introduction to such methods.

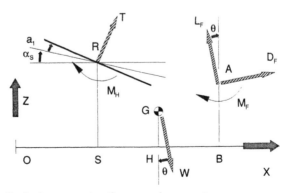

**Fig. 9.3.** Longitudinal trim geometry, forces and moments.

*Coordinates of reference points.* Rotor hub: $x_r$ (*OS*) and $z_r$ (*SR*). Centre of gravity coordinates: $x_g$ (*OH*) and $z_g$ (*HG*). Fuselage aerodynamic origin: $x_a$ (*OB*) and $z_a$ (*BA*), and also $\theta$ is the fuselage pitch attitude, $\alpha_s$ is the main rotor shaft tilt and $a_1$ is the rearward disc tilt relative to shaft axes.

Resolving parallel to *OB*

$$T\sin(\alpha_s + a_1) + W\sin\theta + D_F\cos\theta - L_F\sin\theta = 0$$
$$T\sin(\alpha_s + a_1) = (L_F - W)\sin\theta - D_F\cos\theta$$

(9.6)

Resolving normal to *OB*

$$T\cos(\alpha_s + a_1) + L_F\cos\theta + D_F\sin\theta - W\cos\theta = 0$$
$$T\cos(\alpha_s + a_1) = -(L_F - W)\cos\theta - D_F\sin\theta$$

(9.7)

From which the following results are obtained

$$T = \sqrt{(L_F - W)^2 + D_F^2}$$

$$\tan(\alpha_s + a_1) = \frac{(L_F - W)\sin\theta - D_F\cos\theta}{-(L_F - W)\cos\theta - D_F\sin\theta}$$

$$= \frac{\dfrac{D_F}{(L_F - W)} - \tan\theta}{1 + \dfrac{D_F}{(L_F - W)}\cdot\tan\theta}$$

(9.8)

If $\theta_F$ is the backward tilt of the rotor thrust force, then

$$\tan\theta_F = \frac{D_F}{L_F - W}$$

(9.9)

We have from (9.8) and (9.9)

$$\alpha_s + a_1 = \theta_F - \theta$$
$$a_1 = \theta_F - \theta - \alpha_s$$

(9.10)

To evaluate the fuselage attitude $\theta$ we use the moment balance equation.
   Taking clockwise moments about *R*

$$W\cos\theta\cdot(x_G - x_R) - W\sin\theta\cdot(z_R - z_G) - (L_F\cos\theta + D_F\sin\theta)\cdot(x_A - x_R)$$
$$-(-L_F\sin\theta + D_F\cos\theta)\cdot(z_R - z_A) + M_s\cdot a_1 + M_F = 0$$

(9.11)

which simplifies to

$$A\cos\theta + B\sin\theta + M_S \cdot a_1 + M_F = 0 \tag{9.12}$$

where

$$\begin{aligned}
A &= W(x_G - x_R) - L_F(x_A - x_R) - D_F(z_R - z_A) \\
B &= -W(z_R - z_G) - D_F(x_A - x_R) + L_F(z_R - z_A)
\end{aligned} \tag{9.13}$$

Substituting for $a_1$ from Eqn (9.10) we find

$$\begin{aligned}
A\cos\theta + B\sin\theta + M_S(\theta_F - \theta - \alpha_s) + M_F &= 0 \\
A\cos\theta + B\sin\theta - M_S\theta + [M_S\theta_F - M_S\alpha_s + M_F] &= 0
\end{aligned} \tag{9.14}$$

This is a transcendental equation in $\theta$, which can be solved very readily by the Newton–Raphson technique, for example.

The solution for $\theta$ unlocks the calculation and $a_1$ can now be determined.

The fuselage and rotor disc attitude are now determined. To complete the analysis the flapping and thrust equations from Chapter 6 can now be used to evaluate the rotor control angles required together with the rotor coning angle.

Recalling Eqn (6.98)

$$\begin{bmatrix} \lambda_\beta^2 & 0 & 0 & 0 \\ \dfrac{\gamma}{6}\mu_x & 1-\lambda_\beta^2 & -\dfrac{\gamma}{8} & 0 \\ 0 & \dfrac{\gamma}{8} & 1-\lambda_\beta^2 & 0 \\ 0 & 0 & 0 & 1 \end{bmatrix} \cdot \begin{bmatrix} a_0 \\ a_1 \\ b_1 \\ \dfrac{C_T}{s}\cdot\dfrac{1}{a} \end{bmatrix} = \begin{bmatrix} \dfrac{\gamma}{8} & 0 & -\dfrac{\gamma}{6}\mu_x & -\dfrac{\gamma}{6} \\ 0 & -\dfrac{\gamma}{8} & 0 & 0 \\ \dfrac{\gamma}{3}\mu_x & 0 & -\dfrac{\gamma}{8} & 0 \\ \dfrac{1}{3} & 0 & -\dfrac{1}{2}\mu_x & -\dfrac{1}{2} \end{bmatrix} \cdot \begin{bmatrix} \theta_0 \\ A_1 \\ B_1 \\ \mu_z + \lambda_i \end{bmatrix} \tag{9.15}$$

Rearranging gives

$$\begin{bmatrix} \dfrac{\gamma}{8} & 0 & -\dfrac{\gamma}{6}\mu_x & -\lambda_\beta^2 \\ 0 & -\dfrac{\gamma}{8} & 0 & -\dfrac{\gamma}{6}\mu_x \\ \dfrac{\gamma}{3}\mu_x & 0 & -\dfrac{\gamma}{8} & 0 \\ \dfrac{1}{3} & 0 & -\dfrac{\mu_x}{2} & 0 \end{bmatrix} \cdot \begin{bmatrix} \theta_0 \\ A_1 \\ B_1 \\ a_0 \end{bmatrix} = \begin{bmatrix} \dfrac{\gamma}{6}\mu_{zD} \\ (1-\lambda_\beta^2)a_1 - \dfrac{\gamma}{8}b_1 \\ \dfrac{\gamma}{8}a_1 + (1-\lambda_\beta^2)b_1 \\ \dfrac{C_T}{sa} + \dfrac{\mu_{zD}}{2} \end{bmatrix} \tag{9.16}$$

from which the rotor control and coning angles are determined.

The final solution gives the following data

$$\text{fuselage attitude} = \theta$$

$$\text{blade flapping angles} = a_0, \ a_1, \ b_1$$

$$\text{blade control angles} = \theta_0, \ A_1, \ B_1$$

## Shaft/disc axes

It should be noted that $\mu_x$, $\mu_z$ and $\lambda_i$ for the flapping equations are in shaft axes, but in order to evaluate $\lambda_i$ by a momentum approach disc axes must be used, i.e. for shaft axes

$$\mu_{x\,\text{SHAFT}} = \mu \cos(\alpha_s + \theta)$$
$$\mu_{z\,\text{SHAFT}} = -\mu \sin(\alpha_s + \theta)$$

$\qquad$ (9.17)

and for disc axes

$$\mu_{x\,\text{DISC}} = \mu \cos(\alpha_s + \theta + a_1)$$
$$\mu_{z\,\text{DISC}} = -\mu \sin(\alpha_s + \theta + a_1)$$

$\qquad$ (9.18)

the minus signs are included in the second parts of both equations because in the momentum analysis, $\mu_z$ and $\lambda_i$ are assumed positive downwards.

## Hub moment term $M_S$

The moment contribution of the hub through disc tilt is determined by $M_S$. In order to evaluate this, the following equation allows it to be calculated from the usual specification of blade dynamic data, i.e. blade flapping inertia, rotor speed and fundamental flapping frequency.

To analyse the hub moment the rotor is modelled with zero flapping hinge offset and rigid blades which have a root spring included to adjust the flapping frequency, from which we have [see Eqn (6.8)]

$$\omega_\beta^2 = \Omega^2 \left[ 1 + \frac{k_\beta}{I_\beta \Omega^2} \right]$$
$$k_\beta = I_\beta \Omega^2 (\lambda_\beta^2 - 1)$$

$\qquad$ (9.19)

If the rotor tilts longitudinally by an angle $a_1$, then each blade performs a sinusoidal flapping motion of amplitude $a_1$, giving the spring moment

$$M_{\text{SPRING}} = k_\beta a_1 \cos \psi \tag{9.20}$$

We now require the component of this moment in the pitch direction.
   Resolving into the pitch axis

$$\begin{aligned} \mathrm{d}M_{\text{PITCH}} &= M_{\text{SPRING}} \cos \psi \\ &= k_\beta a_1 \cos^2 \psi \end{aligned} \tag{9.21}$$

then on averaging around the rotor disc, and including all $N$ blades we find

$$\begin{aligned} M_{\text{PITCH}} &= \frac{N}{2\pi} k_\beta a_1 \int_0^{2\pi} \cos^2 \psi \, \mathrm{d}\psi \\ &= \frac{N}{2} k_\beta a_1 \end{aligned} \tag{9.22}$$

then finally

$$M_S = \frac{N}{2} k_\beta \tag{9.23}$$

The control of the rotor on the fuselage can be expressed as the moment derived about the helicopter CG for a unit tilt of the rotor disc. The moment is derived from two sources: (i) the tilt of the rotor disc giving an equal tilt to the thrust vector and moving its line of action away from the CG and (ii) the flapping accompanying the disc tilt will impart a moment directly to the hub through the effect of flapping stiffness. The former source (i) is directly determined from the rotor thrust and the latter (ii) is dependent also on the rotor hub characteristics, notably the flapping frequency.

   In Chapter 6 the natural flapping frequency of a blade was raised above once per revolution, not only by a root spring but also by non-zero flapping hinge offset. This offset also gives rise to a hub moment when the rotor disc is tilted. With reference to Fig. 9.4 the blade tension can be seen to generate the moment due to their lines of action being offset by the flapping hinge positions. Whatever the origin of the flapping frequency it is the determining factor in the generation of hub moment from tilting the rotor disc.

   With reference to Fig. 9.5 consider the effect of longitudinal disc tilt on the pitching moment where the rotor disc is tilted by an angle $a_1$.

   The moment about the CG is then given by

$$\begin{aligned} M &= Th a_1 + M_S a_1 \\ &= a_1 (Th + M_S) \end{aligned} \tag{9.24}$$

from which

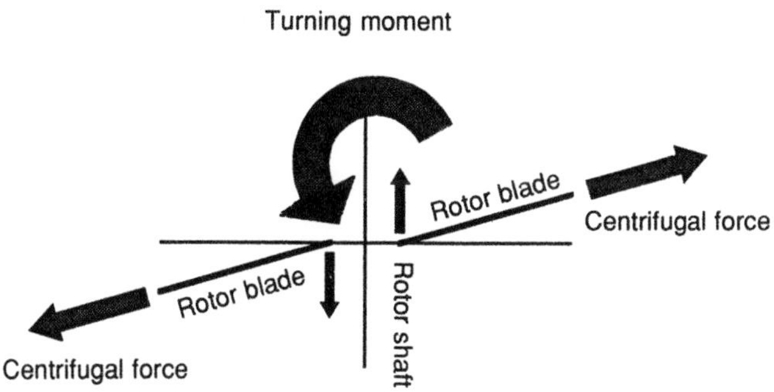

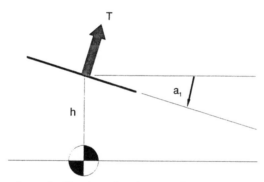

**Fig. 9.4.** Origin of centrifugal stiffness due to offset flapping hinges.

**Fig. 9.5.** Hub moment due to inclination of main rotor thrust.

$$\frac{\mathrm{d}M}{\mathrm{d}a_1} = Th + M_S \qquad (9.25)$$

using Eqns (9.19) and (9.22) this becomes

$$\frac{\mathrm{d}M}{\mathrm{d}a_1} = Th + \frac{N}{2} I_\beta \Omega^2 (\lambda_\beta^2 - 1) \qquad (9.26)$$

If the load factor on the rotor is $n$, then

$$T = nW \qquad (9.27)$$

from which

$$\frac{\mathrm{d}M}{\mathrm{d}a_1} = nWh + \frac{N}{2} I_\beta \Omega^2 (\lambda_\beta^2 - 1) \qquad (9.28)$$

if we normalise this expression on its value when $n = 1$, $\lambda_\beta = 1$ (i.e. a hovering teetering rotor) we have

$$\left(\frac{\mathrm{d}M}{\mathrm{d}a_1}\right)_{\mathrm{REF}} = Wh \tag{9.29}$$

from which

$$\frac{\mathrm{d}M / \mathrm{d}a_1}{(\mathrm{d}M / \mathrm{d}a_1)_{\mathrm{REF}}} = n + \frac{N}{2}\frac{I_\beta \Omega^2}{Wh}(\lambda_\beta^2 - 1) \tag{9.30}$$

The effect of the load factor and flapping frequency are demonstrated in the following example

Number of blades $(N) = 4$<br>
Rotor speed $(\Omega) = 30$ rad/s<br>
Flapping inertia $(I_\beta) = 800$ kg m$^2$<br>
All-up weight $(W) = 50\,000$ Newtons<br>
Height of rotor centre above CG $(h) = 2$ m

The results are plotted in Fig. 9.6 for different flapping frequencies, representing teetering, articulated and semi-rigid rotors. As can be seen, the possibility of transferring high hub moments to the fuselage improves the control and trim behaviour remarkably. The hinge offset or hub stiffness gives a control moment that is independent of the thrust. This is important in manoeuvres when low values of load factor are being used. For very low load factors the articulated rotor loses most of its control power, while the teetering rotor relies totally on the main rotor thrust for its control, and the loss of control power at low or zero rotor thrust can give rise to the catastrophic phenomenon of mast bumping.

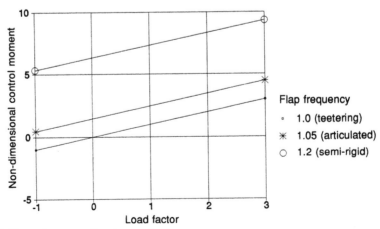

**Fig. 9.6.** Control power of various rotor types.

## Fuselage aerodynamics

The fuselage aerodynamic data $L_F$, $D_F$ and $M_F$ can be evaluated in several ways; the analysis here will use the reference values at a given speed of 100 m/s (i.e. $L_{100}$, $D_{100}$ and $M_{100}$) which are usually obtained from wind tunnel test results.

Assuming that the drag varies with the square of the forward speed we have

$$
\begin{bmatrix} L_F \\ D_F \\ M_F \end{bmatrix} = \frac{\sigma V^2}{10^4} \begin{bmatrix} L_{100} \\ D_{100} \\ M_{100} \end{bmatrix} \tag{9.31}
$$

where $\sigma$ is the relative air density.

It should be emphasised that the following analysis assumes the terms $L_{100}$, $D_{100}$ and $M_{100}$ are invariant with fuselage attitude. This is purely to emphasise the type of calculation that can be used, and to see a simple method in use. To include the full effects of the fuselage and the rotor drag ($H$ force) will require a more involved iteration scheme.

There are many effects of fuselage attitude on the aerodynamics, but to indicate some examples:

1.  $L_{100}$ will be particularly sensitive to wings.
2.  $D_{100}$ is greatly influenced with flow changes over the rear fuselage structure, particularly if a rear loading ramp door is fitted.
3.  $M_{100}$ will change with the fitting of a tailplane. Use of this aerodynamic surface allows the designer more freedom to adjust the trim of the aircraft.

The method described to calculate the trim of a helicopter is simpler than usual in order to show the considerations required in its derivation. A simple method should not be condemned if it helps to gain a feel or understanding of a problem. Simplified analyses have their place but the user must never lose sight of the approximations made and not to infer more information than the simplifications justify.

### Lateral trim

Lateral trim can be analysed in a similar manner, the tail rotor thrust being an important difference.

## Example of the method

The method has been applied to several example cases, the data for which are given in Table 9.1. The basic rotor data used are

rotor tip speed = 200 m/s
rotor radius = 10 m
blade chord = 0.5 m
shaft tilt angle = 0°

**Table 9.1**

| Case | 1 | 2 | 3 |
|---|---|---|---|
| All-up mass (kg) | 10 000 | 10 000 | 10 000 |
| $x_G$ (m) | 0 | 0.5 | 0 |
| $z_G$ (m) | 0 | 0 | 0 |
| $x_R$ (m) | 0 | 0 | 0 |
| $z_R$ (m) | 2.0 | 2.0 | 2.0 |
| $x_A$ (m) | 0 | 0 | 0 |
| $z_A$ (m) | 1.0 | 1.0 | 1.0 |
| $L_{100}$ (N) | 0 | 0 | 0 |
| $D_{100}$ (N) | 12 000 | 12 000 | 12 000 |
| $M_{100}$ (N) | 0 | 0 | 0 |
| No. of blades | 5 | 5 | 5 |
| $\lambda_\beta$ | 1 | 1 | 1.5 |
| $I_\beta$ (kg m²) | 2400 | 2400 | 2400 |

The results are presented in Figs 9.7–9.9 for cases 1–3, respectively.
All three cases show similar trends against forward speed

1. The fuselage attitude rotates progressively forward.
2. The $B_1$ cyclic pitch increases to tilt the rotor forward to overcome the drag of the fuselage.
3. The $A_1$ cyclic pitch gradually increases in magnitude (but of negative sign) to counteract the lateral disc tilt caused by the effect of coning angle. The method will only show the $A_1$ cyclic pitch required in order to trim out any lateral disc tilt. In practice a small amount of lateral disc tilt will be required to balance the tail rotor thrust. This will itself vary with the torque requirement of the main rotor. This is a good illustration of the strong coupling between longitudinal and lateral trim effects.

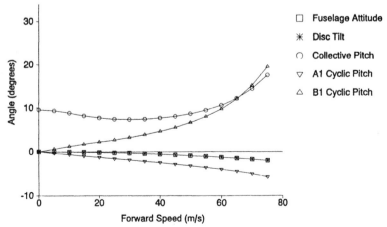

**Fig. 9.7.** Longitudinal trim equilibrium—case 1.

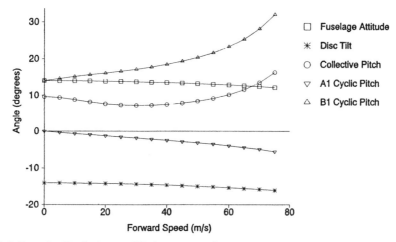

**Fig. 9.8.** Longitudinal trim equilibrium—case 2.

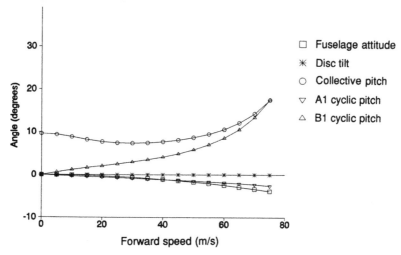

**Fig. 9.9.** Longitudinal trim equilibrium—case 3.

4.  The collective pitch decreases from its hover value to a minimum at a moderate forward flight speed. This is the benefit of the translational velocity component on the lifting capacity of the rotor. As the forward flight speed increases further, the increasingly nose-down attitude of the fuselage causes an increased downflow $(\mu_z)$ which must be counteracted by increasing the collective pitch.

5.  The differences between cases 1 and 2 are the position of the aircraft CG. Case 1 has a neutral CG position whilst case 2 has an aft CG position where the fuselage attitude is more nose-up, requiring an increase in $B_1$ cyclic pitch to achieve the required forward disc tilt. The 0.5 m CG position is very severe but serves to illustrate its effect.

6.  Cases 1 and 2 are for a teetering rotor $(\lambda_\beta = 1)$, case 3 has a very stiff rotor $(\lambda_\beta = 1.5)$ and the disc tilt is reduced from approximately 2.5° to essentially zero. This is the effect of the high control power afforded by a stiff hub. As

will be discussed in Chapter 10, the provision of high control power carries the disadvantage of providing a better route for vibration to be transmitted from the rotor hub to the fuselage.

## Incidence vs Mach number plots

Having established the trim variables of the helicopter for a given weight and forward flight speed, it is now possible to monitor the blade behaviour in more detail. This is achieved by calculating the incidence and Mach number seen by an individual blade element as it rotates about the azimuth. These are sometimes colloquially called "sausage plots" because of their shape, but they are of use in the determination of where possible rotor limits may be crossed over the entire rotor disc and where stall regions can be expected. The analysis conducted so far can now be used to perform this task.

Selecting a non-dimensional radial station of $x$ at an azimuth of $\psi$, the non-dimensional vertical and tangential velocities are given by

$$\frac{U_P}{V_T} = \lambda_i + \mu_z + \mu_x \beta \cos \psi + x \beta'$$

$$\frac{U_T}{V_T} = x + \mu_x \sin \psi$$

(9.32)

The flapping angle is given by

$$\beta = a_0 - a_1 \cos \psi - b_1 \sin \psi$$

(9.33)

This gives the non-dimensional flapping velocity as

$$\beta' = a_1 \sin \psi - b_1 \cos \psi$$

(9.34)

The blade pitch angle is

$$\theta = \theta_0 - A_1 \cos \psi - B_1 \sin \psi$$

(9.35)

The inflow angle is

$$\phi = \tan^{-1}\left(\frac{U_P}{U_T}\right)$$

(9.36)

The incidence angle is now as follows:

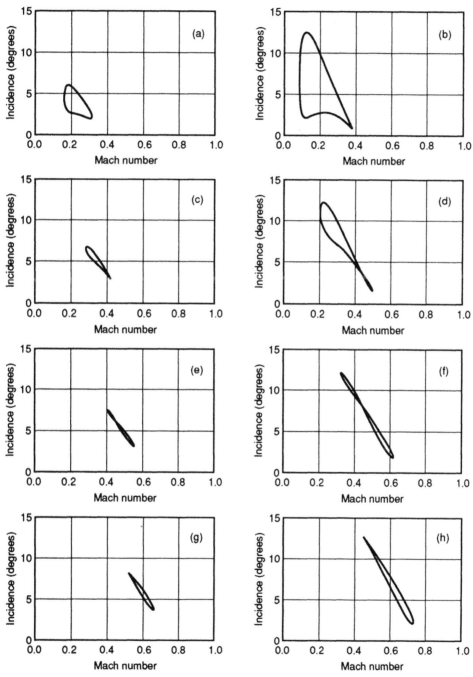

**Fig. 9.10(a–h).** (a) Radial station 40%, forward speed 25 m/s. (b) Radial station 40%, forward speed 50 m/s. (c) Radial station 60%, forward speed 25 m/s. (d) Radial station 60%, forward speed 50 m/s. (e) Radial station 80%, forward speed 25 m/s. (f) Radial station 80%, forward speed 50 m/s. (g) Radial station 100%, forward speed 25 m/s. (h) Radial station 100%, forward speed 50 m/s.

$$\alpha = \theta - \phi \tag{9.37}$$

The Mach number is given by

$$M = M_T(x + \mu_x \sin \psi) \tag{9.38}$$

where $M_T$ is the blade tip Mach number.

The aircraft configuration of case 1 gives the following incidence vs Mach number plots for forward speeds of 25 and 50 m/s at radial stations of 40%, 60%, 80% and 100%. These are presented in Figs 9.10(a)–(h).

As can be seen from the figures, the variation between the plots is quite marked. The blade stations near the root are subject to high incidences at low Mach numbers and are thus prone to high incidence stall. Blade stations near the tip suffer from compressibility effects as can be seen from the plots but the incidence values are reduced compared to the innermost parts of the blade. It must be emphasised that these calculations were made assuming the blade is untwisted, and putting blade twist terms in the equations will cause changes in the loops. The effects of this can readily be accommodated in the analysis by including a $\kappa x$ term in the blade pitch expression.

# 10

# Vibration and its transmission

## Origins of the forcing

The helicopter relies on the aerodynamic forces produced by the rotor(s) to fly. Since the environment surrounding the blades changes greatly during their rotation around their respective shafts, much variation in the aerodynamic loads will be encountered. Therefore the root cause of vibration in a helicopter is that a rotor blade is subjected to varying factors which greatly influence the aerodynamic forces acting on the blades. The effects can be listed as

(a) Variations of dynamic pressure due to the combination of the rotational velocity and any forward velocity.
(b) Changes in pitch angle due to cyclic pitch.
(c) Variations in the inflow angles because of blade flapping or lagging motion.
(d) The effects of vortex wake interaction with the blades.

It is to be appreciated that the rotor blade experiences many features of an edgewise translating rotor which results in variations of the aerodynamic loads placed upon it. Each individual blade will experience this forcing and when the contribution of each blade is added together to give the overall effect on the rotor hub, some important characteristics are found. This chapter shows how these characteristics arise and how the helicopter designer handles this problem of vibration and its transmission to the rest of the airframe.

## Transmission of the blade forcings

Consider a rotor consisting of $N$ blades, where each blade is subjected to a forcing of $n$ per revolution (frequency $= n\Omega$), of magnitude $A_n$. The force can be expressed as

$$L(\psi) = A_n \cos n\psi \qquad (10.1)$$

In order to specify the rotor azimuthal position, one blade, the primary blade, is set at an azimuth of $\psi$, and the remaining blades are equally spaced around the rotor with an inter-blade spacing angle of $\phi$, where

$$\phi = \frac{2\pi}{N} \qquad (10.2)$$

Hence the $k$th blade is positioned at an azimuth angle of $\psi_k$ given by

$$\psi_k = \psi + (k-1)\phi \qquad (10.3)$$

then the overall forcing for $N$ blades is given by

$$F = \sum_{k=1}^{N} \sum_{n=0}^{\infty} A_n \cos n\psi_k \qquad (10.4)$$

This assumes a direct transmission of the force from the blade to the rotor hub, e.g. the vertical bounce transferred via the vertical shear forces to the hub. Substituting (10.3) into (10.4) we have

$$F = \sum_{n=0}^{\infty} A_n \left[ \sum_{k=1}^{N} \cos n(\psi + \overline{k-1}\,\phi) \right] \qquad (10.5)$$

This expression and its equivalent sine form are summations which can readily be derived by using the result

$$e^{in\theta} = \cos n\theta + i \sin n\theta \qquad (10.6)$$

If we now define the summations

$$C = \sum_{k=1}^{N} \cos n(\psi + \overline{k-1}\,\phi)$$

$$S = \sum_{k=1}^{N} \sin n(\psi + \overline{k-1}\,\phi) \qquad (10.7)$$

we can combine them into one using

$$C + iS = \sum_{k=1}^{N} e^{in(\psi + \overline{k-1}\phi)}$$

$$= e^{in\psi} \cdot \sum_{k=1}^{N} e^{in\overline{k-1}\phi} \qquad (10.8)$$

The summation is a geometric progression with the first term $= 1$, a multiplying factor of $e^{in\phi}$ and $N$ as the number of terms. The sum of the progression is then

$$C + iS = e^{in\psi} \left[ \frac{(e^{in\phi})^N - 1}{e^{in\phi} - 1} \right] \qquad (10.9)$$

In view of the expression (10.2) for the inter-blade spacing we find

$$(e^{in\phi})^N = e^{inN\phi}$$
$$= e^{2\pi in} \tag{10.10}$$
$$= 1$$

Hence

$$C + iS = 0 \tag{10.11}$$

which at first sight seems to suggest that all forcings cancel out, leaving none at the rotor head. This is patently false as the case of $n = 0$ shows, corresponding to steady lift. Since the numerator of (10.9) must always be zero, the only way that a non-zero value of $C + iS$ can be obtained is when the denominator of (10.9) is also zero. For this to be true

$$e^{in\phi} = 1$$
$$\Leftrightarrow in\phi = (2\pi i)m$$
$$\Leftrightarrow in\frac{2\pi}{N} = 2\pi im \tag{10.12}$$
$$\Leftrightarrow n = mN$$

where $m$ is an arbitrary integer.

In other words, for any possible chance of a non-zero result, the forcing frequency expressed relative to the rotor speed must be an integer multiple of the number of blades. In this case the evaluation of the expression (10.9) requires a limit procedure since the right-hand side is of the form

$$\frac{0}{0} \tag{10.13}$$

If $n$ is regarded as a general non-integer number and approaches a value of $mN$ as a limit, then L'Hopital's rule can be invoked, giving

$$C + iS = \lim_{n \to mN} \left[ \frac{\frac{\partial}{\partial n}(e^{inN\phi} - 1)}{\frac{\partial}{\partial n}(e^{in\phi} - 1)} \right]$$
$$= \lim_{n \to mN} \left[ \frac{iN\phi e^{inN\phi}}{i\phi e^{in\phi}} \right] \tag{10.14}$$
$$= N$$

Summarising

$$C + iS = Ne^{in\psi}, \quad \text{if } n = mN$$
$$= 0, \quad \text{otherwise}$$

(10.15)

Finally we have the overall effect of the individual blade forcings in combination as

$$F = \sum_{n=0}^{\infty} (A_n \cos n\psi) \cdot \delta_{n,mN}$$
$$= N \sum_{m=0}^{\infty} A_{mN} \cos mN\psi$$

(10.16)

where $\delta_{ij}$ is the Kronecker delta defined by

$$\delta_{ij} = 1, \quad i = j$$
$$= 0, \quad i \neq j$$

(10.17)

As an example of how the forcings from the blades of an $N$ bladed rotor cancel or reinforce, consider a rotor with the primary blade at azimuth $\psi = 0°$. A forcing of unit magnitude but frequency of $n$/rev. (cos $n\psi$) is shown for all the blades at

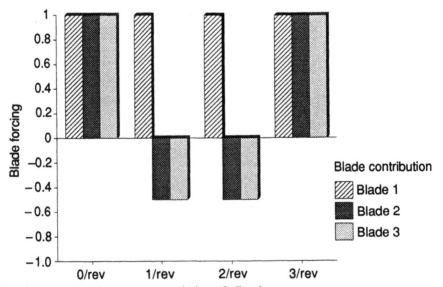

**Fig. 10.1.** Three-bladed rotor: transmission of vibration.

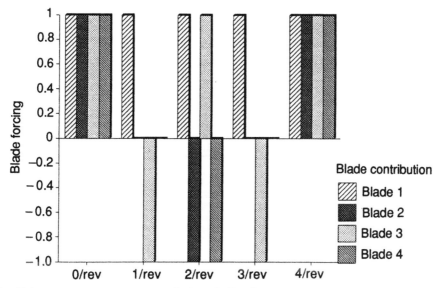

**Fig. 10.2.** Four-bladed rotor: transmission of vibration.

their respective azimuth locations, together with the total effect. The pattern, as previously derived, is shown for $N = 3, 4, 5$ in Figs 10.1–10.3, and is tabulated in Tables 10.1–10.3.

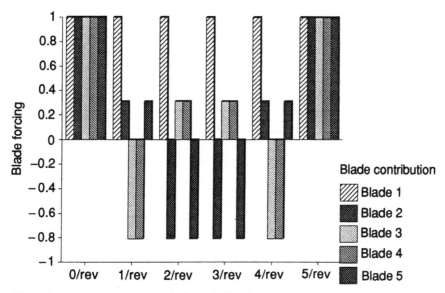

**Fig. 10.3.** Five-bladed rotor: transmission of vibration.

**Table 10.1.** $N = 3$

| Forcing frequency | $\psi = 0°$ | $\psi = 120°$ | $\psi = 240°$ | Total |
|---|---|---|---|---|
| 0 | 1 | 1 | 1 | 3 |
| 1 | 1 | −0.5 | −0.5 | 0 |
| 2 | 1 | −0.5 | −0.5 | 0 |
| 3 | 1 | 1 | 1 | 3 |

**Table 10.2.** $N = 4$

| Forcing frequency | $\psi = 0°$ | $\psi = 90°$ | $\psi = 180°$ | $\psi = 270°$ | Total |
|---|---|---|---|---|---|
| 0 | 1 | 1 | 1 | 1 | 4 |
| 1 | 1 | 0 | −1 | 0 | 0 |
| 2 | 1 | −1 | 1 | −1 | 0 |
| 3 | 1 | 0 | −1 | 0 | 0 |
| 4 | 1 | 1 | 1 | 1 | 4 |

**Table 10.3.** $N = 5$

| Forcing frequency | $\psi = 0°$ | $\psi = 72°$ | $\psi = 144°$ | $\psi = 216°$ | $\psi = 288°$ | Total |
|---|---|---|---|---|---|---|
| 0 | 1 | 1 | 1 | 1 | 1 | 5 |
| 1 | 1 | 0.31 | −0.81 | −0.81 | 0.31 | 0 |
| 2 | 1 | −0.81 | 0.31 | 0.31 | −0.81 | 0 |
| 3 | 1 | −0.81 | 0.31 | 0.31 | −0.81 | 0 |
| 4 | 1 | 0.31 | −0.81 | −0.81 | 0.31 | 0 |
| 5 | 1 | 1 | 1 | 1 | 1 | 5 |

It can be seen that a rotor filters out all direct forcing frequencies except multiples of the number of blades. This has the advantage of restricting the amount of vibration transmitted, but has the danger of removing the ability to observe a build up of blade loadings, should an instability occur at a frequency where cancellation occurs.

Not all forcings are directly transformed in the same way as the vertical bounce. This vertical forcing, for instance, will also excite a roll motion as can be seen in Fig. 10.4. Since the lift force for most of the azimuth rotation will be offset from the helicopter centreline a rolling moment will be created.

In this situation, we have

$$M_{\text{ROLL}}(\psi) = L(\psi) \cdot r \sin\psi$$

$$= \sum_{n=0}^{\infty} A_n \cos n\psi \cdot r \sin\psi \qquad (10.18)$$

$$= \frac{r}{2}\left[\sum_{n=0}^{\infty} A_n (\sin\overline{n+1}\,\psi - \sin\overline{n-1}\,\psi)\right]$$

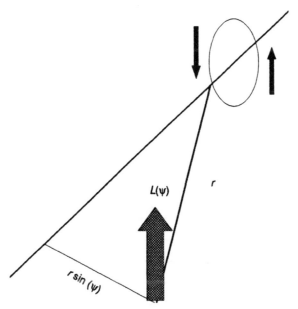

**Fig. 10.4.** Vertical forcing giving roll moment.

These forcings will be summed over the $N$ blades and from the previous analysis, a cancellation will occur unless

$$n \pm 1 = mN$$
$$\Leftrightarrow n = mN \pm 1 \tag{10.19}$$

then the roll moment will have contributions from forcings whose frequency relative to the rotor speed is equal to a multiple of the number of blades, plus or minus one. However, it is important to realise that these forcing frequencies are in the rotor frame of reference. In the fuselage they will be felt at a frequency of $mN$.

For example, a four-bladed rotor will have

$$0, 4, 8, \ldots \text{ per revolution—via the vertical bounce}$$

or

$$1, 3, 5, 7, 9, \ldots \text{ per revolution—via the roll moment.}$$

but will appear with frequencies of

$$0, 4, 8, \ldots \text{ per revolution}$$

Table 10.4 shows the transmitted forcings for up to 10 blades and 10/rev. forcing frequency. The table is symmetric so either labelling applies.

**Table 10.4**

| | 0 | 1 | 2 | 3 | 4 | 5 | 6 | 7 | 8 | 9 | 10 |
|---|---|---|---|---|---|---|---|---|---|---|---|
| 0 | | | | | | | | | | | |
| 1 | | | | | | | | | | | |
| 2 | | | | | | | | | | | |
| 3 | | | | | | | | | | | |
| 4 | | | | | | | | | | | |
| 5 | | | | | | | | | | | |
| 6 | | | | | | | | | | | |
| 7 | | | | | | | | | | | |
| 8 | | | | | | | | | | | |
| 9 | | | | | | | | | | | |
| 10 | | | | | | | | | | | |

## Excitation within the fuselage

The fuselage has many different vibration modes, all of which may be excited by rotor vibration. If the frequency of the mode happens to coincide with that of a force input from the rotor, then high vibration levels are likely to be experienced in the fuselage. Depending upon the mode excited, the amplitude of the motion produced will vary with the position in the fuselage. There are a number of solutions which can be considered:

(a) Change the frequency and/or damping of that fuselage mode.
(b) Change the number of rotor blades—very useful if a particular aerodynamic exciting component is strong. Not surprisingly, this is only possible in the initial design stages.
(c) Absorb the energy at the rotor head or at a point in the fuselage.
(d) Isolate the vibration.
(e) Superimpose a set of forces to cancel out the response of the fuselage.

## Other sources of vibration

In addition to the basic variation in aerodynamic loads on the rotor blades induced by the forward flight speed and the blade flapping motion, other phenomena contribute to the vibration of a helicopter. Three examples[17] illustrate these phenomena:

- *Pitch–lag instability*. If a blade lags rearwards and at the same time the pitch decreases (due to the geometry of the rotor hinges or elastic deformation) so that the lift decreases, then the rotor will experience an oscillation in the flapwise forces. If the phase of this lag–pitch motion and the blade flapping couple together, the rotor tip path plane will not be steady and this will be perceived in the airframe as a stirring motion. Normally this is a limit cycle. It can also be the first indication of a lag damper whose performance is degrading.

- *Classical flutter*. This is a coupling between torsion and bending of the blades brought about by the chordwise position of the blade CG and is not unlike wing flutter. If the blade twists to increase the incidence (nose up) due to aerodynamic forces, the twist induces additional aerodynamic force. This extra aerodynamic lift causes vertical bending of the blade. The effect is best avoided and if the phase between the torsion and the bending oscillation is correct, then the oscillation will grow. It is usual that the blade is mass balanced, when the centre of mass is on or ahead of the quarter chord position.

- *Stall flutter*. This is a single degree of freedom flutter where the unsteady aerodynamic effects of stall can couple with the pitching motion of the blade section. It can be triggered when the blade is twisted elastically by some mechanism, particularly the aerodynamic pitching moment, in a condition where the aerofoil section is near to its stalling incidence. The blade reacts to the loads and increases the incidence until the section stalls. The stall will give a rapid change to the aerodynamic loads and pitching moment and the lift collapses and the blade untwists. This does not always result in a steadily increasing amplitude oscillation however, but can continue in the form of a limit cycle. The mechanism of the blade section stalling as the incidence increases is different from that of the flow reattachment as the incidence subsequently decreases. (As the blade stalls and the leading edge suction peak collapses, a vortex forms at the leading edge and traverses the upper surface of the aerofoil section. While it is close to the upper surface, lift is generated, not unlike the lift generated by a delta wing at high incidence, which will influence greatly the lift and pitching moment behaviour. Later in the cycle, with the pitch reduced, as the flow reattaches to the upper surface there is now no vortex flow and the effects of the reattachment will be different to those of the stall.) There is a hysteresis and the phasing of the pitching moment with the incidence and pitch rate can induce energy to be drawn from the airstream, thereby maintaining the torsional motion or, at worst, cause a divergent behaviour.

# Dynamic vibration absorber

The dynamic vibration absorber is a simple mass/spring device which is rigidly attached to a forced vibrating system with the aim of suppressing the motion of the parent system. The theoretical model of the parent/absorber combination is shown in Fig. 10.5. Since it has a dynamic response of its own, it can vibrate and the aim of attaching this device is to arrange the response of the absorber to give a forcing to the parent system directly out of phase with the external forcing. In this way the

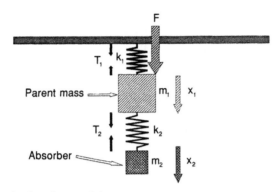

**Fig. 10.5.** Dynamic absorber model.

motion of the original system can be arrested. The parent system is of mass $m_1$ attached to earth by a spring of rate $k_1$. This is subjected to a force $F$. Attached to the mass is the absorber which consists of a mass $m_2$ and a spring of rate $k_2$. The inertial displacement of the parent mass is $x_1$ and that of the absorber $x_2$.

The spring tension loads are given by

$$T_1 = k_1 x_1$$
$$T_2 = k_2(x_2 - x_1)$$

(10.20)

The equations of motion for the masses $m_1$ and $m_2$ are respectively

$$m_1 \ddot{x}_1 = F + T_2 - T_1$$
$$= F + k_2(x_2 - x_1) - k_1 x_1$$
$$m_2 \ddot{x}_2 = -T_2$$
$$= -k_2(x_2 - x_1)$$

(10.21)

Upon rearranging, the final equations are

$$m_1 \ddot{x}_1 + (k_1 + k_2)x_1 - k_2 x_2 = F$$
$$m_2 \ddot{x}_2 + k_2 x_2 - k_2 x_1 = 0$$

(10.22)

Now defining the external forcing, $F$, to be sinusoidal of frequency $\omega$, a solution at this frequency is now sought by substituting

$$F = F_0 \sin \omega t$$
$$x_1 = X_1 \sin \omega t$$
$$x_2 = X_2 \sin \omega t$$

(10.23)

The resulting equations in the response amplitudes $X_1$ and $X_2$ are

$$\begin{bmatrix} k_1 + k_2 - m_1\omega^2 & -k_2 \\ -k_2 & k_2 - m_2\omega^2 \end{bmatrix}\begin{bmatrix} X_1 \\ X_2 \end{bmatrix} = \begin{bmatrix} F_0 \\ 0 \end{bmatrix} \tag{10.24}$$

The solution of which is

$$\frac{1}{F_0}\begin{bmatrix} X_1 \\ X_2 \end{bmatrix} = \frac{1}{(k_1 + k_2 - m_1\omega^2)(k_2 - m_2\omega^2) - k_2^2}\begin{bmatrix} k_2 - m_2\omega^2 \\ k_2 \end{bmatrix} \tag{10.25}$$

It is more instructive to define the spring rates $k_1$ and $k_2$ via the frequencies $\omega_1$ and $\omega_2$ where

$$k_1 = m_1\omega_1^2$$
$$k_2 = m_2\omega_2^2 \tag{10.26}$$

$\omega_1$ and $\omega_2$ are the natural frequencies of the parent/absorber mass/spring combinations in isolation, respectively.

The final solution becomes

$$\frac{1}{F_0}\begin{bmatrix} X_1 \\ X_2 \end{bmatrix} = \frac{1}{[m_1(\omega_1^2 - \omega^2)(\omega_2^2 - \omega^2) - m_2\,\omega^2\omega_2^2]}\begin{bmatrix} \omega_2^2 - \omega^2 \\ \omega_2^2 \end{bmatrix} \tag{10.27}$$

The above results are shown in Fig. 10.6 for the following system

$$\omega_1 = 5 \text{ rad/s}, \qquad \omega_2 = 10 \text{ rad/s}$$
$$m_1 = 4 \text{ kg}, \qquad m_2 = 1 \text{ kg}$$
$$F_0 = 4 \text{ N}$$

From these results it is apparent that the absorber will function with maximum efficiency when

$$\omega = \omega_2 \tag{10.28}$$

with the responses

$$\frac{1}{F_0}\begin{bmatrix} X_{1\,min} \\ X_{2\,min} \end{bmatrix} = \begin{bmatrix} 0 \\ -\dfrac{1}{m_2\omega_2^2} \end{bmatrix}$$

$$= \begin{bmatrix} 0 \\ -\dfrac{1}{k_2} \end{bmatrix} \tag{10.29}$$

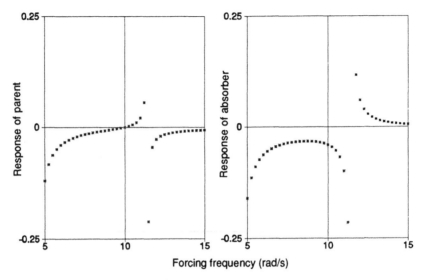

Forcing frequency (rad/s)

**Fig. 10.6.** Dynamic absorber model responses.

(i.e. when the amplitude of the parent mass motion falls to zero). This result also shows that the amplitude of the absorber response can be minimised by using a stiff spring which, in order to give the correct frequency, will require a large mass, with the accompanying weight problems.

## Passive vibration absorption

The dynamic vibration absorber is a passive device and some examples of its application to absorb rotor head and fuselage vibration are described below.

### Dynamic absorber (rotor head)—Westland Lynx[18]

The Westland Lynx head absorber is shown in Figs 10.7, 10.8 and Plate 6.1. In essence it consists of a central spindle which is bolted to the hub centre in place of the lifting eye, which is normally used for hub removal during maintenance. The outer ring is attached to the spindle by means of four composite leaf springs. There is a top and bottom cover fitted for protection against the elements and also to allow tuning weights to be fitted, allowing frequency adjustment. The outer ring, top and bottom covers, and the tuning weights form the mass $m_2$ and the four composite leaf springs provide an axisymmetric spring $k_2$. It is of a fixed frequency $\omega_2$ and therefore will rapidly lose its effect if the rotor speed, and hence the forcing frequency $\omega_1$, changes during helicopter operation. In the analysis of the dynamic vibration absorber no damping effects were included. In reality a vibration absorber will possess a degree of internal damping which causes two

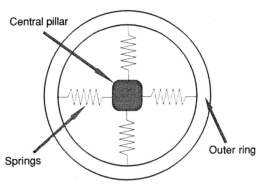

**Fig. 10.7.** Westland head absorber (plan view).

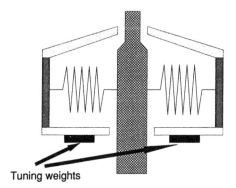

**Fig. 10.8.** Westland head absorber (side view).

effects. Firstly the efficiency of the absorber will be reduced, but secondly the range of rotor speed over which the absorber is effective will widen. Therefore an amount of rotor speed variation can be accommodated which is within that provided by a modern engine/rotor governor system. The design of a vibration absorber will have to balance efficiency with range of operating frequency. Usually, the rotor speed is held virtually constant by means of an engine governor system, enabling a fixed frequency vibration absorber to function effectively.

### Battery dynamic absorber (fuselage)—Westland Sea King

Situated in the nose of the Westland Sea King is a battery fitted to the fuselage by means of a sprung mount, as shown in Fig. 10.9. The mass of the battery and spring mount characteristics are tuned to absorb the dominant main rotor vibration forces. It is also of fixed natural frequency $\omega_2$ and, like the dynamic rotor head absorber, will detune as the forcing frequency varies.

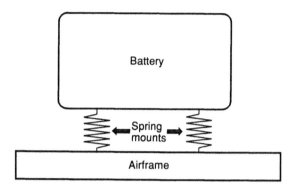

**Fig. 10.9.** Sea King battery absorber.

## *Bifilar dynamic absorber (rotor head)—Sikorsky Blackhawk, S92*

The bifilar absorber is shown in Fig. 10.10. Rather than a mass and spring to provide the vibrating mechanism for the vibration absorber, a pendulum in a centrifugal force field will provide the same effect. The bifilar, which is a variation on a pendulum absorber, consists of a weight mounted by means of two rolling pivot pins to a bracket which is itself fixed rigidly to the rotor head. The whole assembly lies in the plane normal to the rotor shaft. As the weight moves from side to side, the pins rolling in the circular holes in the rotating bracket cause the weight itself to move like a pendulum towards the hub centre. The weight actually moves in a circular path (as would a simple pendulum) whose radius is equal to half of the difference in the diameters of the holes and the pins. Hence the effective length of the pendulum can be adjusted by the sizing of the components. Since the dynamics of the bifilar are identical to a pendulum, the characteristics will be the same. It has its own natural frequency which will depend on the pendulum length and the centrifugal force field. For small oscillations this will vary directly with the rotor speed and therefore will automatically retune as the rotor speed varies during operation. Should considerable weight movement occur then (exactly like a simple pendulum), the bifilar absorber will not perform an exact simple harmonic motion. If this happens the natural bifilar frequency will not remain locked precisely to the rotor speed and the absorber will not remain precisely in tune, however it has been used successfully. The advantage of the bifilar is that it provides a compact means of giving a short pendulum length without concern to the size of the weights employed. However, in use it can suffer from both wear and the effects of the operating environment which means that maintenance will be required.

## Passive vibration isolation

The vibration absorber is used to reduce the vibration in a system by opposing the forcing. The vibration isolator, however, stands between the source of the vibration and the system with intention of confining the vibration to an area and hence

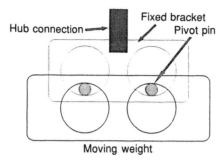

**Fig. 10.10.** Bifilar dynamic absorber.

isolate the majority of the fuselage from the source of vibration. It is usually a passive device.

A method of reducing the vibration transmitted from the rotor to the fuselage is to suspend the fuselage by spring units underneath the rotor and transmission. The dynamic equations of motion are a special case of the dynamic absorber with $k_1 = 0$. The response of the fuselage is therefore $X_2$ and is given by

$$\frac{X_2}{F_0} = \frac{-1}{m_2\omega^2 + m_1\omega^2\left(1 - \dfrac{\omega^2}{\omega_2^2}\right)} \qquad (10.30)$$

In the static case, the fuselage will simply hang under the rotor system with a deflection $d$ where

$$\begin{aligned} m_2 g &= k_2 d \\ &= m_2\omega_2^2 d \end{aligned} \qquad (10.31)$$

from which

$$\omega_2^2 = \frac{g}{d} \qquad (10.32)$$

then

$$\frac{X_2}{F_0} = \frac{-1}{m_2\omega^2 + m_1\omega^2\left(1 - \omega^2\dfrac{d}{g}\right)} \qquad (10.33)$$

If we define a transmission ratio, $T_R$, as the ratio of the fuselage vibration amplitude to that of the rigid case where $d = 0$, we have

$$T_R = \frac{m_1 + m_2}{m_2 + m_1\left(1 - \omega^2\dfrac{d}{g}\right)} \qquad (10.34)$$

If the fuselage mass ratio $M$ is defined as

$$M = \frac{m_1}{m_2} \qquad (10.35)$$

we have for the transmission ratio

$$T_R = \frac{M + 1}{1 + M\left(1 - \omega^2\dfrac{d}{g}\right)} \qquad (10.36)$$

which on rearranging to isolate the static deflection, $d$, gives

$$d = \frac{g}{\omega^2}\left[1 - \frac{1}{M}\left(\frac{M+1}{T_R} - 1\right)\right] \qquad (10.37)$$

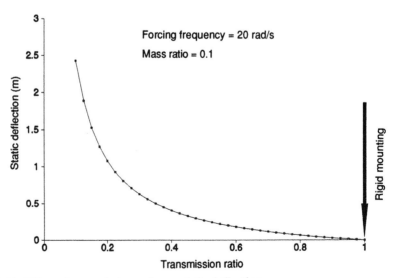

Forcing frequency = 20 rad/s

Mass ratio = 0.1

**Fig. 10.11.** Effect of transmission ratio on static deflection.

If a representative example is analysed where $M = 0.1$ and $\omega = 20$ rad/s the static deflection magnitude for a given transmission ratio, $T_R$, is presented in Fig. 10.11. As can be seen, the reduction in transmission of vibration from the rotor to the fuselage is accompanied by a large static deflection of the suspension. Good isolation is therefore achieved with a very "soft" mounting. A helicopter during its life will usually have to make some extreme manoeuvres and the idea of the fuselage being mounted to the gearbox with highly flexible couplings is not realistic. This is plainly unacceptable, particularly considering possible agile manoeuvres of the helicopter. The concept of a simple fuselage mount as a vibration isolator requires a compromise. A more sophisticated device must be used to achieve a sensible mount which securely attaches the fuselage to the rotor and transmission but provides a high degree of vibration isolation.

## Dynamic anti-resonant vibration isolator (DAVI)[19]

One such method of obtaining a favourable vibration isolation of the fuselage, without unacceptable deflection of the suspension, is that of the dynamic anti-resonant vibration isolator (DAVI) which puts a pendulum device in parallel with the spring mount. The DAVI isolator is shown schematically in Fig. 10.12 and is a device used to support the fuselage underneath a framework to which is installed the gearbox and engines. A possible layout is shown in Fig. 10.13. The DAVI is a combination of a spring and a pendulum device which when correctly tuned prevents the transmission of vibration from the framework to the fuselage. The swinging mass and the spring effectively cancel each other out and the fuselage, experiencing the summation of these two effects will not receive any forcing and hence will not suffer vibration.

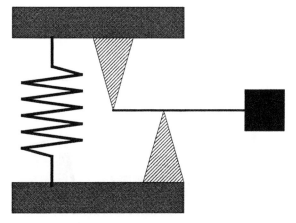

**Fig. 10.12.** Dynamic anti-vibrator isolator (DAVI).

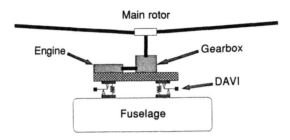

**Fig. 10.13.** DAVI-based isolator.

## *Nodal suspension—Bell Nodamatic™* [20]

The nodal suspension system (see Fig. 10.14) is similar to the DAVI system in that the fuselage is suspended under the gearbox assembly. However, in this case the method of attachment is a tuned beam. The forcing applied to the beam by the main rotor will cause it to vibrate in a manner whereby two nodes are seen either side of the attachment of the beam to the transmission. The fuselage is mounted underneath this beam at the nodes and in this way the vibration is not transmitted to the fuselage.

# Active vibration cancellation

The five systems described above are all of a passive nature in operation. In recent times, active anti-vibration schemes have been developed to mimic the principle of a vibration isolator only to supply the negative cancelling forces by means of actively controlled actuators rather than the dynamic behaviour of a passive isolator.

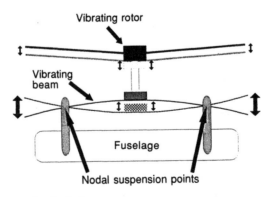

**Fig. 10.14.** Nodal suspension isolation system.

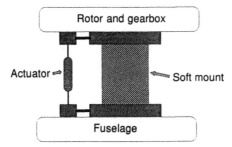

**Fig. 10.15.** Active vibration cancellation system—Westland ACSR.

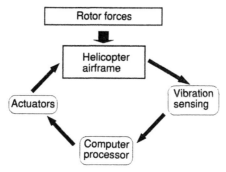

**Fig. 10.16.** ACSR flow diagram.

## *Westland ACSR system*[21]

The mechanism of the Westland Active Control of Structural Response (ACSR) system is shown in Fig. 10.15. It was used originally with the Westland 30 aircraft which has a soft mount isolation system fitted as standard and actuators were fitted in parallel with the mounts. The installation is similar to that shown in Fig. 10.13. The original mounts were retained alongside the actuators for reasons of safety should an actuator suffer a mechanical failure. Sensors, such as accelerometers, were mounted to the airframe at critical points and the measured responses from these transducers were compared with a specified vibration level. Differences were used as inputs to an algorithm which caused the actuators to adjust their forcing in order to cancel out the response of the fuselage. The flow diagram is shown in Fig. 10.16. The sampling is not continuous, but in discrete units of time, and the actuator behaviour updated accordingly.

# Tail rotors

## Miscellaneous points of interest

We have earlier observed that the vast majority of helicopters are of the single main and tail rotor configuration. The main rotor provides control for five of the six degrees of freedom: *surge, sway, heave, roll* and *pitch*. The final degree of freedom, *yaw*, is controlled by the tail rotor, whose only responsibility is to provide a sideways thrust force and thereby produce a yawing moment about the main rotor shaft. At first sight this seems a less than onerous task, but when this requirement is set in the context of the location of the tail rotor and the flight, and therefore aerodynamic, environment in which it is expected to perform, it puts a different slant on the duties of the tail rotor. Additionally, if the consequences of the dynamic behaviour of the helicopter on the tail rotor are observed, then the tasks required of the tail rotor become even more difficult.

This chapter aims to highlight the requirements of a tail rotor, its problems, and how these can be overcome.

To set the scene several general observations are now made.

1. Tail rotor radius is approximately one sixth of that of the main rotor. (A brief summary of rotor sizes of past and present helicopters is presented in Table 11.1 and shown in Fig. 11.1.)
2. Tail rotor tip speed is approximately equal to that of the main rotor. (In view of noise regulations, the limitation of tail rotor tip speed is becoming of increasing importance.)
3. The tail rotor achieves torque balance with a power consumption of the order of 10% of that consumed by the main rotor.
4. The tail rotor provides yaw accelerations of the order of 1 rad/s$^2$.

**Table 11.1**

| Aircraft | All-up weight (kg) | Main rotor diameter/ tail rotor diameter |
|---|---|---|
| Robinson R 22 | 623 | 7.19 |
| Sikorsky S64 | 19 091 | 4.51 |
| McDonnell Douglas AH64 | 9526 | 5.24 |
| Westland Sea King | 9318 | 6.0 |
| Mil-Mi 26 | 49 500 | 4.2 |

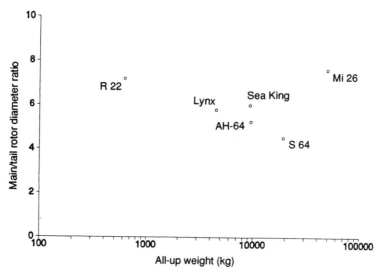

**Fig. 11.1.** Comparative sizing of main and tail rotors.

5.  The weathercock stiffness of a tail rotor is equal to a circular disc of the same
    area at approximately 40 knots. The figure of 40 knots is relevant if the lift
    curve slope of the disc is taken to be the same as for the tail rotor blades. This
    means that the disc is regarded as of high aspect ratio. To be more realistic
    perhaps, the circular disc should be modelled by slender wing theory in which
    case equality of weathercock stiffness is achieved at a speed of about 140
    knots.
    Take for instance a helicopter, in forward flight, whose main rotor rotates an-
    ticlockwise when viewed from above. The tail rotor will then be thrusting to
    starboard. Should the fuselage slew nose left say, then the forward speed will
    have a component along the tail rotor shaft effectively superimposing a climb
    velocity. With a fixed collective pitch this will cause a reduction in the tail
    rotor thrust in which case the yawing moment of the tail rotor will not be
    sufficient to balance that of the main rotor causing the fuselage to move nose
    right and restore the direction of the helicopter. If the nose slews nose right,
    the tail rotor now has a descent velocity superimposed, giving an increase in
    the tail rotor thrust. The tail rotor torque is now larger than that of the main
    rotor and the fuselage moves nose left, again restoring the helicopter's direc-
    tion. In other words, the tail rotor provides a weathercock stability by creating
    the appropriate restoring moment for a yaw excursion.
6.  The tail rotor is sometimes the noisiest component (especially if two-
    bladed). The noise energy produced by a tail rotor does not normally exceed
    that of the main rotor, however, since the tail rotor has a much greater
    rotational speed, the noise energy is at a frequency more discernible to the
    human ear.
7.  In the majority of cases the tail rotor is located close to the main rotor height to
    avoid undue roll coupling in hover, which places it in the main rotor hub

**Plate 11.1.** Westland Sea King main rotor head with blades fully folded. Note the dog house fairing and beanie cap.

turbulent wake, in forward flight. To minimise the adverse effects on the tail rotor, fuselage modifications have been used to limit this effect (the dog house/ horse collar fairing behind the main rotor hub or the beanie cap placed above the main rotor hub being two familiar examples, see Plate 11.1).

# Aerodynamics of tail rotors

It is assumed that the direction of rotation of the main rotor is counterclockwise when viewed from above. This is the convention used in British or American designs. French and Russian helicopter designs usually have main rotors which rotate in the opposite clockwise direction.

In designing a tail rotor many requirements have to be considered. The low disc loading tail rotor is by far the most efficient approach to torque compensation and directional control for the single main rotor helicopter. However, it is a very difficult task to develop a tail rotor that has completely acceptable control, stability and structural characteristics. The tail rotor operates in an extremely adverse aerodynamic and dynamic environment and must be capable of producing thrust with the free stream coming from any direction. It is mounted on a fin/boom which through aerodynamic interference causes a loss in thrust. Additional aerodynamic interference is produced by the main rotor wake and when the aircraft is hovering close to the ground, the effect of ambient wind direction on this interference is a major consideration in the performance of the tail rotor.

Tests[22] have been conducted on the velocities impinging on the tail rotor region under the influence of the main rotor hub and fairings. The results are shown

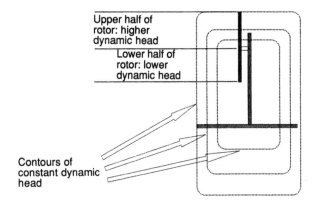

Upper half of
rotor: higher
dynamic head

Lower half of
rotor: lower
dynamic head

Contours of
constant dynamic
head

**Fig. 11.2.** Aerodynamic interference at the tail rotor.

diagrammatically in Fig. 11.2 as contours of constant velocity. It can be seen that the top half of the tail rotor experiences a higher dynamic head than the lower half which will affect the tail rotor in a similar manner to the effect that forward speed has on the main rotor. This effect can be balanced out by arranging the tail rotor to rotate in a sense where the top blade moves aft. The subject of the rotational direction has much importance on tail rotor performance and is discussed later in this chapter.

The tail rotor differs from the main rotor in that it is not required to be trimmed by using cyclic pitch which is therefore not fitted. It must also be able to produce both positive and negative thrusts. The only control provided is through foot pedals which alter the tail rotor collective pitch.

The effect of altitude and temperature on the hovering helicopter will need to be addressed regarding the tail rotor performance and the engine power available. The manoeuvrability of the helicopter will have to be considered as will the ability of the aircraft to operate in high winds from any direction.

A final consideration is that posed by the inevitability of later developments to the original helicopter design which can be of an increased performance or the ability to operate in more adverse atmospheric conditions. It is an unwritten law that "an aircraft's weight always seems to increase with development". A somewhat hard nosed statement, but based on experience. As the aircraft weight increases, so does the main rotor power demands, which, in turn, puts greater requirements on the tail rotor. The phrase "out of sight, out of mind" seems appropriate to the tail rotor, however, it deserves as much consideration as the main rotor.

## Maximum thrust

In all flight regimes the tail rotor must produce sufficient thrust to counteract any adverse aerodynamic effects of the fin, to react the main rotor torque, to manoeuvre the helicopter in yaw and to correct for any disturbances. In cruise the tail rotor thrust can be relieved by a cambered fin providing a side force and offloading the tail rotor. However, this can cause a problem in autorotation since in this condition the main rotor is externally driven by the air and there is no need for torque reaction. The fin force will remain and yaw trim will require a reverse force to be applied by the tail rotor. Normally, the low-speed thrust requirements determine a tail rotor's design, but the thrust required in manoeuvres at high speed and consequently at high tip Mach numbers should always be checked. The maximum tail rotor thrust is generally established from a consideration of the critical maximum sidewards flight velocity in hover or at low speeds, and the desired manoeuvrability in yaw at near zero speed. These thrust requirements when combined with the thrust needed to counteract the main rotor torque, determine the maximum thrust.

## Precession

During a low-speed yaw manoeuvre the tail rotor thrust must compensate for the main rotor torque, accelerate the aircraft in yaw and accommodate the effects of tail rotor precession at the yaw rate of the aircraft. Compared to the main rotor, the tail rotor will be subject to appreciably larger rates of rotation about an axis in the plane of rotation, with spot turns being the most frequent occurrences. This causes the gyroscopic effects of the rotor to be felt, particularly in the aerodynamic performance. The dynamics of rotating bodies show that in order to cause a rotation of the angular velocity vector of a spinning body, a moment must be applied 90° in advance, as shown in Fig. 11.3.

With a rolling/pitching moment, in order for the tail rotor disc to follow the shaft, the blades will require such a moment, but cannot achieve it via the flapping hinge. It can only be achieved with blade flapping inducing a difference in lift between two sides of the rotor disc which has the effect of causing the rotor plane to follow the rotor shaft but lag behind by an angle dependent on the rotation rate of the rotor shaft. Figures 11.4–11.6 show the moment, flapping velocity and flapping behaviour necessary to generate this moment. As the rotor disc is inclined to the rotor shaft, the rotor thrust will be similarly orientated during such motions. In the case of the main rotor, this has a beneficial effect as it causes a thrust-depend-

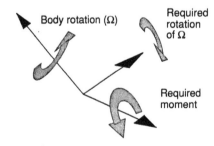

**Fig. 11.3.** Gyroscopic precession.

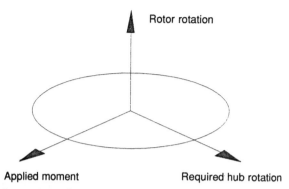

**Fig. 11.4.** Precession and required moment.

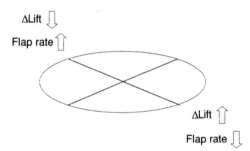

**Fig. 11.5.** Precession and required flapping velocity.

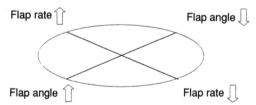

**Fig. 11.6.** Precession and required flapping displacement.

ent moment to be generated about the CG should the airframe roll or pitch. As the rotation rate of the shaft increases, so does the disc tilt and hence the moment for the rotor thrust. The moment acts in a direction against the motion and is proportional to its rate; because of this it behaves as a damper. The tail rotor, however, finds this effect detrimental as it drives half of the rotor harder due to the flapping increasing the blade loading and conversely offloading the other half. This has the immediate problem of pushing some blade sections toward stall.

The angle the rotor disc lags behind the shaft is given by

$$\frac{16}{\gamma}\frac{q}{\Omega} \tag{11.1}$$

where $\gamma$ is the blade Lock number, $q$ is the disc rotation angular velocity and $\Omega$ is the rotor speed, i.e. the angle by which the rotor disc lags behind the shaft increases with a reduction in Lock number, which is a by-product of increasing the blade weight. Therefore a large blade weight not only causes design difficulties because of an increase in weight at the end of the tail boom, but also causes part of the tail rotor to be driven harder towards stall.

If stall is encountered, then the moment generated by the loaded side is limited. If an increase in rotation rate is still required an additional precessional moment from the unstalled side of the rotor disc must be produced which can only be achieved with a reduction in the rotor thrust. The thrust capability of the tail rotor is therefore significantly reduced. See Fig. 11.7, which is taken from Lynn *et al.*[23]

After subtracting the thrust required to overcome the main rotor torque, the stall boundary can be plotted as a function of yaw rate and yaw acceleration[23] (see Fig. 11.8). Large flapping angles occur as the unstalled side of the rotor disc attempts to create all of the required precessional moment. Stall is most likely to occur when the rotor thrust and yaw rate are both large, i.e. stopping a nose right hovering turn.

In forward flight the situation is somewhat different. Although yaw rates may be lower than in hovering manoeuvres, the precession has the effect of increasing the retreating blade incidence when the helicopter yaws nose left (this is independent of the direction of rotation of the tail rotor), and stall may therefore occur in forward flight when turning left. The onset of precessional stall may be delayed by using an aerofoil with an increased $C_{L\,MAX}$,[23] an increased blade Lock number, or by increasing the tip speed.

Suggested design criteria[23] are to start a left hovering turn with an initial accel-

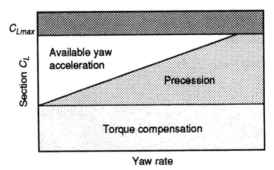

**Fig. 11.7.** Effect of precession on yaw acceleration.

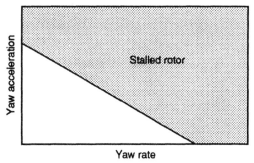

**Fig. 11.8.** Yaw performance boundary.

eration of 1 rad/s², or to stop a right hovering turn rate of 0.75 rad/s with an initial deceleration of 0.4 rad/s². The first is critical for the thrust and the second for the precessional moment required for blades of low Lock number. These manoeuvres are performed at the critical ambient design condition.

## Design considerations

In addition to providing a mounting point for the tail rotor, a vertical fin can offload the tail rotor in cruise and hence reduce the rotor flapping. Unfortunately, the fin affects the low speed yaw characteristics by significantly reducing the net thrust and increasing the power requirements. These considerations must be included in the maximum thrust requirements. Fin interference is a function of the fin size and position in relation to the tail rotor, the flight speed and direction, and the main rotor wake. There are two types of tail rotor installation, tractor and pusher, which are considered below.

### Tractor

In this installation, shown in Fig. 11.9, the hovering wake strikes the fin and the rotor thrust is directly away from it. The resulting side-force generated by the fin subtracts from the tail rotor thrust and the overall net thrust is given approximately by

$$T_{net} = T\left(1 - 0.75\frac{S}{A}\right)$$    (11.2)

where

$$\frac{S}{A} = \frac{\text{blocked disc area}}{\text{total disc area}}$$    (11.3)

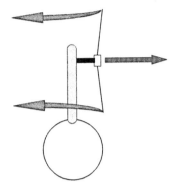

**Fig. 11.9.** Tractor tail rotor.

These results are described in Lynn *et al.*[23]

To put this in perspective a 20% reduction in the net tail rotor thrust reduces the yaw acceleration capability of a helicopter by about 60%.

## *Pusher*

In this installation, as shown in Fig. 11.10, the tail rotor thrusts directly at the fin and hence sucks air over the fin surface nearest to the rotor. This causes an area of lower pressure to form on this side of the fin with the result of a force being generated which acts in opposition to the tail rotor thrust. In addition, there is an efficiency loss due to the distorted inflow. The wake will be essentially undisturbed by the fin (see Plate 11.2).

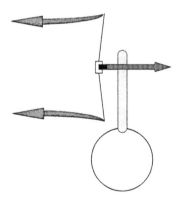

**Fig. 11.10.** Pusher tail rotor.

**Plate 11.2.** EH101 prototype PP5 tail rotor and fin. (This is a pusher tail rotor.) Note the pitch change mechanism and the proximity of the rotor plane to the fin.

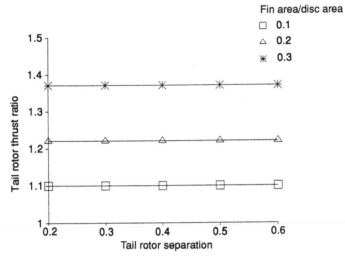

**Fig. 11.11.** Fin/tail rotor interference—tractor.

## Fin interference

The fin side-force depends on the fin size and position relative to the tail rotor, wind velocity and main rotor wake. With reference to Figs 11.11 and 11.12, which are from Stepniewski and Keys,[13] the effects of the fin/rotor separation distance and the blocked disc area are shown for the tractor and pusher, respectively. The ordinate is the ratio of the gross tail rotor thrust and the net value with the interfering fin force subtracted. As the interference increases, the ratio also increases from the isolated rotor case value of unity. As can be seen, the fin/rotor separation distance is a major parameter for the pusher, unlike the tractor where it has virtually

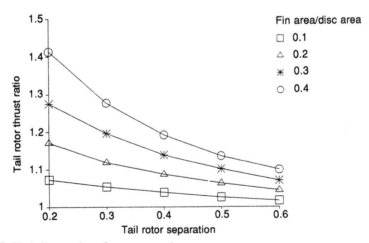

**Fig. 11.12.** Fin/tail rotor interference—pusher.

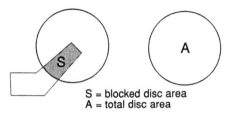

S = blocked disc area
A = total disc area

**Fig. 11.13.** Tail rotor disc area blocked by the fin.

no effect. For a separation distance greater than 80% of the rotor radius, the fin force is negligible in the case of the pusher. The fin area/disc area parameter is the fraction of the tail rotor disc which is covered by the fin (Fig. 11.13).

# Disc loading

Among the various considerations in determining the tail rotor diameter[23] and hence disc loading are the helicopter size and the effect of the tail rotor power required and weight on the overall helicopter performance. Small changes in total power required which could be obtained by careful attention to the tail rotor design can result in a weight saving or an increase in payload. A typical power saving of 2%, say, results in a payload gain of about 12%. Typical values of tail rotor disc loading when counteracting the main rotor torque are of the order of 400 N/m², but can double that value during critical manoeuvres.

In addition to the tail rotor diameter, the rotor tip speed and number of blades need to be established and noise, profile power, blade stall, rotor torque, weight and control forces are all factors which need to be considered. A rotor with a high tip speed will be generally lighter, require a lower torque and is less susceptible to stall at high speed and during yaw manoeuvres, when compared with a low tip speed design. However, the profile power and the noise will be higher and in recent times noise is becoming a vital factor in the success of a particular helicopter's acceptability.

Negative twist or washout on the tail rotor blade has been used to improve the spanwise loading. In hover and low-speed forward flight, blade twist is helpful in reducing the torque at high thrust. Figure 11.14 shows a typical blade loading distribution with an aerofoil section stall boundary. The proximity of the loading peak to the boundary can be alleviated by using twist and aligning the loading with the stall boundary.[22] However, because the inflow can be from either side of the rotor disc, blade twist is not advantageous for the tail rotor and may affect the blade aerodynamic pitching moments. Twist can improve the maximum thrust capability of the tail rotor but as the twist increases so does the abruptness of the stall behaviour. For low speed helicopters, twist is considered because of the higher hovering efficiency. These are typified by the crane helicopter which requires efficient performance in the hover with little requirement for agile yaw manoeuvres.

Choice of the correct aerofoil section for a tail rotor blade is important because apart from a light blade design (delays precessional stall) and an increase in tip

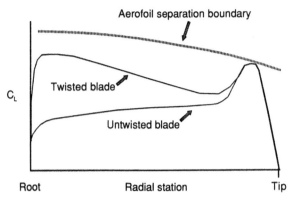

**Fig. 11.14.** Effect of twist on blade loading.

speed, it is the only method of minimising the adverse characteristics of a high thrust design (high torque, increased weight). The main feature is a high $C_{L\,MAX}$ at the operational Mach number and Reynolds number,[24] see Fig. 11.15. Low $C_D$ and aerodynamic pitching moments tend to be a secondary consideration. Aerofoil sections developed specifically for tail rotor blades can suffer from high values of pitching moment which, if ignored, will cause constant heavy loads to the operating servojacks, or with a hydraulic pressure failure, very high loads on the pedals requiring considerable effort on the part of the pilot. Use of preponderance weights, which are described later, allow the designer to adjust the pitching moments to close to zero. It is unwise to trim all of the pitching moment to zero since in the case of tail rotor control failure or breakage there is no mechanism to define the tail rotor blade incidence, with the consequent absence of control. A residual aerodynamic pitching moment is advantageous to the blades as the moment can be used to set essentially zero tail rotor thrust, allowing an autorotative descent to be achievable.

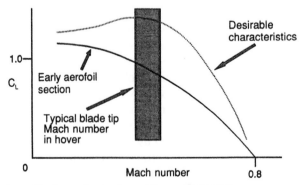

**Fig. 11.15.** Tail rotor—ideal aerodynamic section performance.

# Control

## *Pitch range*

The directional handling qualities of a helicopter greatly depend on the correct tail rotor pitch range and the associated rudder pedal travel or gearing. The pitch range will have to encompass flight conditions varying from high speed right sideways flight requiring high pitch[24] to manoeuvring flight in autorotation requiring low pitch.[23]

Byham[22] suggests the following values for a modern naval helicopter

40%    pitch range to cover tail rotor thrust range in hover ~ 16°

25%    pitch range to give control in autorotation ~ 10°

35%    pitch range to deal with sideways flight inflow ~ 14°

Total ~ 40°

The yaw rate sensitivity is important to the handling. Yaw rate sensitivity is a measure of the yaw acceleration obtained per unit pedal travel. However, the acceptable pedal travel is ±75 mm, and so with the pedal and the pitch travel both fixed the designer has little freedom to adjust the yaw rate sensitivity.

# Sideways flight

Major aerodynamic problems for a tail rotor occur in sideways flight. Since tail rotor performance directly affects the overall yawing moment, these difficulties relate to the directional control.

In left sideways flight, difficulty in establishing pedal trim is experienced at around 5–20 knots as the tail rotor enters the vortex ring state (see Fig. 11.16). This is directly analogous to the main rotor vortex ring state in descending flight

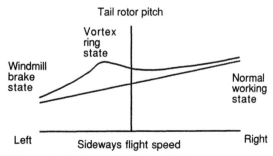

**Fig. 11.16.** Effect of sideward velocity on tail rotor pitch.

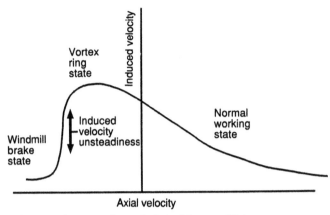

**Fig. 11.17.** Rotor downwash unsteadiness in left sideways flight.

discussed in Chapter 4. The resulting strong vortex formation increases the rotor power required and has the effect of producing a non-uniform flow through the rotor. The vortex ring state is encountered at a higher sideways speed due to the increased downwash obtained with higher values of disc loading compared to the main rotor. Also, in left sideways flight, as can be seen in Fig. 11.17, the average downwash undergoes large changes and so small changes in the incident wind speed can cause abrupt changes in the downwash. This will cause an unsteadiness in the tail rotor thrust which will need to be trimmed out by the pilot adjusting the pedals. As the rotor is in descent there should be few problems as regards power, but there will be an increase in workload for the pilot.

Although the fin blockage can be reduced by increasing the fin/tail rotor separation distance, there will still be a blockage effect in right sideways flight. It is highly desirable to reduce the fin blockage contribution to the tail rotor thrust requirement in order to diminish low-speed flight handling problems.

In sideways flight the lateral velocity and main rotor downwash over the fuselage can combine to cause the tail boom to act as an aerofoil which generates a force in the opposite direction to the tail rotor thrust. A tail boom strake[25] is used on some aircraft to avoid this adverse effect. The strake is positioned to destroy the flow causing the lift and cancel the adverse force. Figure 11.18 shows the principle and Plate 11.3 shows an example.

## Direction of rotation

Figure 11.19 shows representative contours of constant pedal remaining (expressed as a percentage) for a forward-at-the-top rotating tail rotor with an incident wind direction between nose on and starboard of up to 35 knots velocity. The existence of a 0% contour shows that a wind of bearing Green 60°, i.e. at a heading of 60° coming from the starboard side, poses a significant problem. With the wind from that direction the main rotor downwash impinges on the tail rotor (see Figs 11.20 and 11.21). As the main rotor wake moves away from the rotor it distorts under its

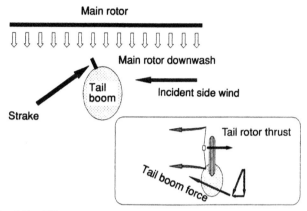

**Fig. 11.18.** Westland Sea King tail boom strake.

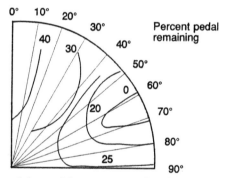

**Fig. 11.19.** Remaining pedal travel for manoeuvre in wind.

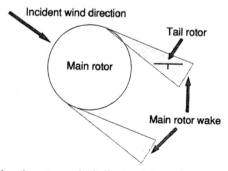

**Fig. 11.20.** Plan view of main rotor wake/tail rotor interaction.

**Plate 11.3.** The tail boom of the EH101 prototype PP5 showing the strake.

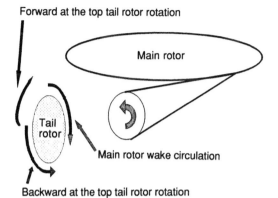

**Fig. 11.21.** General view of main rotor wake/tail rotor interaction.

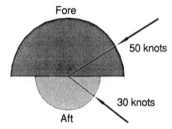

**Fig. 11.22.** Design envelope for tail rotor.

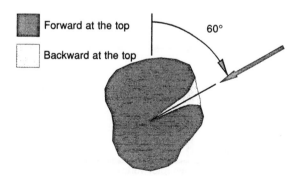

**Fig. 11.23.** Tail rotor envelope with main rotor wake effects.

own influence. This distortion is characterised by each side rolling up into two large circulating areas, rotating from outside to inside, and not unlike a mature fixed wing wake in character. The direction of the rolling up imparts a forward-at-the-top airflow rotation around the tail rotor which will subtract from the dynamic head felt by the rotor blades, because of tail rotor rotation, and thus degrades its thrust producing potential. Figures 11.22 and 11.23 show the effect of main rotor interference on a tail rotor thrust performance. Figure 11.22 is a typical design envelope for helicopter operations. It is a polar plot showing that winds up to a specified limit of, say, 50 knots from the front half of the rotor were intended to be handled. (A lower wind of, say, 30 knots is specified from the rear half.) An example of the type of envelope which can be achieved in practice is shown in Fig. 11.23 and the loss of performance with a Green 60° wind direction is clearly shown. The severe degradation of control can be recovered by reversing the tail rotor direction and filling in this lobe, as shown. For this reason, backward-at-the-top rotation is usually preferred for tail rotors, as can be seen in most airframes. This effect can also be felt when hovering close to the ground in a tail wind,[26] or close to a building. The downwash from the main rotor will spill outwards along the ground and that part moving rearwards past the tail rotor will be influenced by these effects which will redirect the wake upwards allowing it to be reingested into the main rotor and hence form a circulatory flow with forward-at-the-top rotation.

## Burble noise

In addition to giving performance problems, main rotor wake impacts with the tail rotor can generate a significant amount of noise.[27] As the main rotor leaves its vortex wake streaming behind it, the tail rotor will follow and the blades will cut through the wake which is the source of the noise. Forwards-at-the-top rotation usually gives a more periodic nature to these intersections which causes a characteristic noise to be generated. Figure 11.24 shows schematically the timewise distribution of tail rotor blade/main rotor wake intersections which are a series of bursts. The spacing of the bursts is determined by the main rotor blade passages

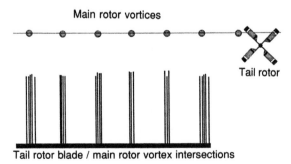

**Fig. 11.24.** Westland Lynx burble noise generation.

leaving the vortices, and the spacing of the impacts within each burst reflects the tail rotor blade passing frequency.

# Yaw control (variations)

In addition to the conventional open tail rotor, various other types of yaw control devices have been used and some examples are highlighted below.

### Fenestron

This is where the tail rotor becomes a fan, and is installed in a duct within a cambered fin to improve efficiency.[28] The thrust/power relationship is the same as an unducted rotor of 30% larger radius. Other benefits include the effects of minimal fin blockage. The advantage of enclosing a fan in a duct has already been discussed in Chapter 4. It has the advantage of safety due to an enclosed rotor disc. Attention needs to be addressed to possible separation around the duct in forward flight since it could spoil the airflow through the fan. The cambered fin allows the fan thrust to be reduced in forward flight since the fin now provides a contribution to the main rotor torque reaction.

### Tail boom circulation control

The concept[29] is illustrated in Figs 11.25 and 11.26. Use of blowing through slots in the tail boom and its interaction with the main rotor downwash will produce a side-force and hence the necessary yawing moment. A fan is used to pressurise the boom and as main rotor downwash is used, an automatic coupling between main rotor torque and tail boom side force occurs. Trimming and rapid yaw control is achieved with a jet located at the rear of the tail boom which swivels allowing the pilot to adjust its direction. It also gives improved safety in use, because of no external rotor disc, and also the danger of tree strikes is minimised.

The sideward force on the tail boom will have a small downwash (induced) component which will tend to alter the pitch trim. A notable example of this technique is the McDonnell Douglas NOTAR™ system.

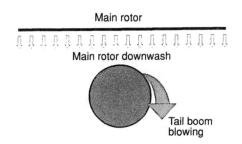

**Fig. 11.25.** The basic principle of tail boom circulation.

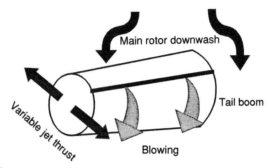

**Fig. 11.26.** NOTAR™—boom blowing and tip jet.

## Canted fin

Canting of the fin[30] provides an upward component of the tail rotor thrust which will help the pitch trim. Additionally, in sideways flight, the tail rotor wake is washed away from the rotor due to the component of sideways flight in the plane of the rotor disc, which will help in the avoidance of the vortex ring state.

## Non-orthogonal blades

A recent battlefield helicopter design (McDonnell Douglas Apache[31]) has a tail rotor with four blades unequally spaced around the azimuth. There are two pairs with a spacing of 55°/125° between adjacent blades. This blade arrangement (see Fig. 11.27), alters the phasing of blade/vortex interactions and is intended to reduce the noise of the tail rotor.

# Propeller moment—preponderance weights

It has previously been mentioned that modern aerofoil sections for tail rotor blades can have large pitching moments. This can cause problems with loads in the control system and indeed the blade pitch behaviour in case of damage to the control linkages. This can be controlled with specially placed weights which make use of the propeller moment.

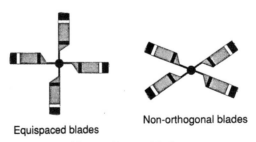

Equispaced blades

Non-orthogonal blades

**Fig. 11.27.** Orthogonal and non-uniform tail rotor blade arrangements.

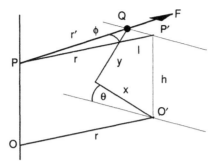

**Fig. 11.28.** Origin of propeller moment.

This is a moment generated by the alignment of the centrifugal forces across the chord of the blade. With reference to Fig. 11.28, consider a blade section located at $O'$, radius $r$ from the centre of the rotor $O$. The blade section rotates in pitch about $OO'$. A general point of the blade section at $Q$ lies $x$ in front of $O'$ and $y$ above it. The blade section is rotated nose up by an angle $\theta$. The centrifugal acceleration felt by the point $Q$ will be in the direction $PQ$ as shown in the figure. If an elemental mass $m$ is positioned at $Q$ then a centrifugal force, $F$, is generated which with reference to the figure has a moment about the pitch axis $OO'$. The following analysis derives the total effect of these moments for the complete blade about $OO'$, which is the propeller moment.

From Fig. 11.28 we see that

$$h = x \sin\theta + y\cos\theta$$
$$l = x \cos\theta - y\sin\theta \tag{11.4}$$

The centrifugal force is given by

$$F = m\,\Omega^2 r' \tag{11.5}$$

The moment of $F$ about the $OO'$ axis is given by

$$M = F\sin\phi \cdot h$$
$$= m\Omega^2 r' \cdot \sin\phi \cdot h \tag{11.6}$$

Noting that

$$r'\sin\phi = l \tag{11.7}$$

We find

$$M = m\Omega^2 \cdot (x\sin\theta + y\cos\theta) \cdot (x\cos\theta - y\sin\theta)$$
$$= m\Omega^2[(x^2 - y^2) \cdot \sin\theta\cos\theta + xy \cdot (\cos^2\theta - \sin^2\theta)] \qquad (11.8)$$
$$= m\Omega^2[(x^2 - y^2) \cdot \tfrac{1}{2}\sin 2\theta + xy \cdot \cos 2\theta]$$

This result is now integrated over the blade giving

$$\text{propeller moment} = \Omega^2\left[\tfrac{1}{2}\sin 2\theta \cdot (I_{xx} - I_{yy}) + \cos 2\theta \cdot I_{xy}\right] \qquad (11.9)$$

where

$$I_{xx} = \int_{\text{BLADE}} mx^2, \quad I_{yy} = \int_{\text{BLADE}} my^2, \quad I_{xy} = \int_{\text{BLADE}} mxy \qquad (11.10)$$

Sometimes the term $I_{xx} - I_{yy}$ is replaced by the polar inertia $I_{xx} + I_{yy}$ which is only applicable if $I_{xx} \gg I_{yy}$. Variation of $I_{xx}$ and $I_{yy}$ allows any amount of propeller moment to be achieved giving the designer freedom to counteract any aerodynamic pitching moments generated by the blades. ($I_{xy} = 0$ for a symmetric blade.)

Preponderance weights are masses placed on rods fitted to the blade cuff and standing out from the plane of the blade. This allows the $I_{xx} - I_{yy}$ term to be adjusted, allowing the high pitching moments caused by modern tail rotor blade aerofoil sections to be offset by the propeller moment and the high loads in the tail rotor control circuit to be minimised. A typical installation is shown in Fig. 11.29.

## Dynamic effects of a tail rotor

The tail rotor, by virtue of it having no lag hinges, possesses a natural lag frequency greater than the rotor speed, i.e. is supercritical, and therefore does not suffer from ground resonance. (This will be discussed in Chapter 12.) The lag forcing felt by the rotor blades must therefore be kept to a minimum. One method commonly used to achieve this is the $\delta_3$ hinge, which has already been described

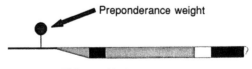

Tail rotor blade

**Fig. 11.29.** Typical location of a preponderance weight.

in Chapter 6. This is a mechanical coupling of the blade pitch to the blade flapping. If the connection between the track rod and the pitch horn does not lie on the flapping hinge line then the kinematics of the linkage will cause any flapping motion to influence the geometric pitch angle on the blade in a sense to arrest the flapping motion. In the case of the main rotor such couplings are avoided by extending the pitch horn towards the rotor shaft, however as the $\delta_3$ hinge provides a passive means of correcting any flapwise blade excursions by adjusting the blade pitch angle, the pitch horn of a tail rotor blade usually extends directly out of the blade cuff to connect with the push rod off the flapping hinge line. (See Plate 6.4.)

It has already been pointed out in Chapter 6 that one effect on the blade behaviour is to raise the natural flapping frequency. This is caused by the influence of the blade aerodynamics and is therefore governed by the blade Lock number.

Chapter 6 showed the effect of $\delta_3$ in the hover. The following analysis examines the influence of $\delta_3$ on the blade flapping in forward flight.

## The effect of $\delta_3$ on tail rotor flapping in forward flight

The analysis of tail rotor blade flapping in forward flight closely follows the analysis of chapter 6 for the main rotor. There are, however, three essential differences

(a)     There is no cyclic pitch applied.
(b)     The rotor shaft remains normal to the flight direction. ($\mu_x = \mu$, $\mu_z = 0$.)
(c)     The $\delta_3$ hinge inserts an extra term in the blade pitch expression.

The blade is assumed to be freely articulated with zero hinge offset.

The tangential and normal velocities incident on a general blade section are given by

$$U_T = V_T(x + \mu \sin \psi)$$
$$U_P = V_T(\lambda_i + \beta'x + \mu\beta\cos\psi)$$

$$(11.11)$$

The blade pitch angle is given by

$$\theta = \theta_0 - D\beta \qquad (11.12)$$

where

$$D = \tan(\delta_3) \qquad (11.13)$$

The aerodynamic lift on a general blade section is

$$L = \frac{1}{2}\rho U_T^2 \, c \, dr \, a \left( \theta - \frac{U_P}{U_T} \right)$$

$$= \frac{1}{2}\rho c \, dr \, a (\theta U_T^2 - U_P U_T) \qquad (11.14)$$

The aerodynamic moment on the blade is thus

$$M_{\text{AERO}} = \frac{1}{2}\rho c a R^2 \int_0^1 (\theta U_T^2 - U_P U_T) x \, dx \qquad (11.15)$$

The centrifugal restoring moment is given by

$$M_{\text{CF}} = -\Omega^2 I_\beta \beta \qquad (11.16)$$

The flapping equation of motion can now be assembled

$$I_\beta \ddot{\beta} = M_{\text{AERO}} - M_{\text{CF}} \qquad (11.17)$$

on substitution of the flapping behaviour

$$\beta = a_0 - a_1 \cos\psi - b_1 \sin\psi$$
$$\beta' = a_1 \sin\psi - b_1 \cos\psi \qquad (11.18)$$
$$\beta'' = a_1 \cos\psi + b_1 \sin\psi$$

the expressions above can be expanded in a Fourier series in $\psi$.

As in the main rotor flapping analysis, the coefficients of the constant $\cos\psi$ and $\sin\psi$ terms are equated giving the following three equations (it should be noted that terms of $O(\mu^2)$ are ignored).

$$\left(1 + \frac{\gamma}{8}D\right)a_0 + \frac{\mu}{3}\left(\frac{\gamma}{4} - 2\right)a_1 - \left(\frac{\mu\gamma}{6}D\right)b_1 - \frac{\gamma}{8}\theta_0 = -\frac{\gamma\lambda_i}{6}$$

$$\left(4\frac{\mu}{3}\right)a_0 - \left(\frac{\gamma}{8}D\right)a_1 - \left(\frac{\gamma}{8}\right)b_1 = 0 \qquad (11.19)$$

$$\left(\frac{\gamma\mu}{3}D\right)a_0 + \left(\frac{\gamma}{8}\right)a_1 - \left(\frac{\gamma}{8}D\right)b_1 - \left(\mu\frac{\gamma}{3}\right)\theta_0 = -\frac{\gamma\mu\lambda_i}{4}$$

To solve the problem we require one more equation which as before comes from the thrust equation

$$T = N\frac{1}{2\pi}\int_0^{2\pi}\left[\frac{1}{2}\rho a\int_0^R c(\theta U_T^2 - U_P U_T)dr\right]d\psi \tag{11.20}$$

Substitution of the above terms gives the following expression

$$\frac{C_T}{sa} = \frac{\theta_0}{3} - \frac{\lambda_i}{2} - a_1\frac{\mu}{2} + b_1\frac{\mu D}{2} - a_0\frac{D}{3} \tag{11.21}$$

These four equations can now be joined to form a $4 \times 4$ matrix equation which can be solved using standard methods

$$
\begin{pmatrix}
1 + \dfrac{\gamma}{8}D & \dfrac{\mu}{3}\left(\dfrac{\gamma}{4} - 2\right) & -\dfrac{\mu\gamma D}{6} & -\dfrac{\gamma}{8} \\[2mm]
4\dfrac{\mu}{3} & -\dfrac{\gamma}{8}D & -\dfrac{\gamma}{8} & 0 \\[2mm]
\dfrac{\gamma\mu D}{3} & \dfrac{\gamma}{8} & -\dfrac{\gamma}{8}D & -\mu\dfrac{\gamma}{3} \\[2mm]
-\dfrac{D}{3} & -\dfrac{\mu}{2} & \mu\dfrac{D}{2} & \dfrac{1}{3}
\end{pmatrix}
\cdot
\begin{pmatrix}
a_0 \\[2mm] a_1 \\[2mm] b_1 \\[2mm] \theta_0
\end{pmatrix}
=
\begin{pmatrix}
-\dfrac{\gamma\lambda_i}{6} \\[2mm]
0 \\[2mm]
-\dfrac{\gamma\mu\lambda_i}{4} \\[2mm]
\dfrac{C_T}{sa} + \dfrac{\lambda_i}{2}
\end{pmatrix}
\tag{11.22}
$$

The fact that the rotor shaft remains normal to the flight direction enables the downwash $\lambda_i$ term to be readily evaluated (i.e. $\mu_z = 0$).

The iterative equation reduces to

$$\lambda_i = \frac{C_T/4}{\sqrt{\mu^2 + \lambda_i^2}} \tag{11.23}$$

which gives the immediate solution

$$\lambda_i = \sqrt{\frac{1}{2}\left[\sqrt{\mu^4 + \frac{C_T^2}{4}} - \mu^2\right]} \tag{11.24}$$

An example of the calculations are shown in Figs 11.30–11.34. This shows the tail rotor blade behaviour for a range of advance ratios and $\delta_3$ angles.

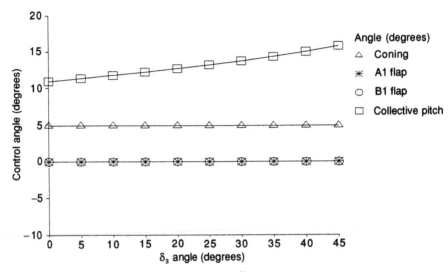

**Fig. 11.30.** Advance ratio = 0.0 for a Sea King tail rotor.

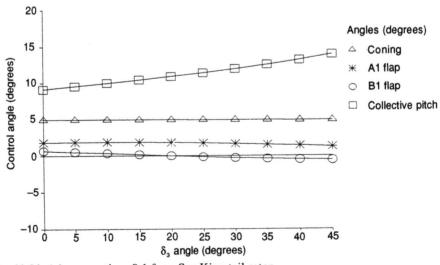

**Fig. 11.31.** Advance ratio = 0.1 for a Sea King tail rotor.

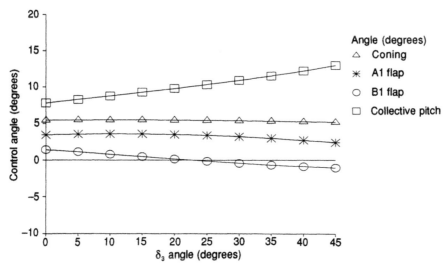

**Fig. 11.32.** Advance ratio = 0.2 for a Sea King tail rotor.

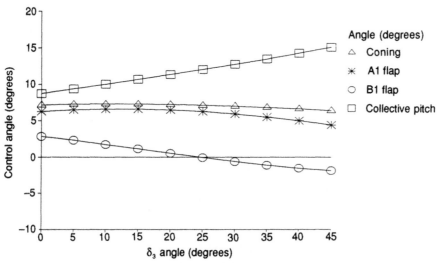

**Fig. 11.33.** Advance ratio = 0.3 for a Sea King tail rotor.

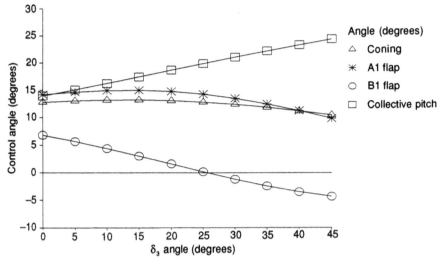

**Fig. 11.34.** Advance ratio = 0.4 for a Sea King tail rotor.

The basic input data are

$$C_T = 0.025$$
$$\text{solidity} = 0.18$$
$$\text{lift curve slope} = 5.8 \, / \, \text{rad}$$
$$\text{Lock number} = 8$$

The solidity is based on the five-bladed Sea King tail rotor. The thrust coefficient $(C_T)$ and Lock number are example values. The effect of forward speed and $\delta_3$ angle on the blade flapping behaviour $(a_0, a_1, b_1)$ and the collective pitch angle $(\theta_0)$ are shown. Figure 11.35 shows the maximum disc tilt angle $(\beta_{MAX})$ for the above cases plotted against the tangent of the $\delta_3$ angle.

Note that

$$\beta_{MAX} = \sqrt{a_1^2 + b_1^2}, \quad \tan \delta_3 = \frac{\partial \theta}{\partial \beta} \qquad (11.25)$$

The reduction in blade flapping with increasing $\delta_3$ angle can be seen. The improvement continues past the value of 45° seen on many helicopters—so the question is posed as to why is the $\delta_3$ angle not increased further? To answer this, one feature of the tail rotor which must be examined is the blade aeroelastic characteristics. Since the $\delta_3$ hinge influences the blade pitch angle, the torsional dynamics of the blade and its control system must be considered. The torsional frequency is governed by the blade torsional inertia and the axial stiffness of the control

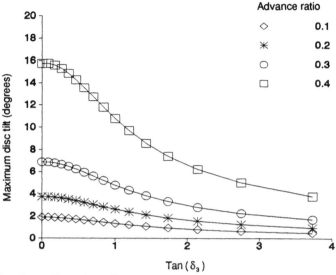

**Fig. 11.35.** Disc tilt—effect of tan $\delta_3$.

system. The latter term has to be treated with caution since the stiffness depends heavily on the type of blade torsional motion. If all the blades rotate in unison about the pitch axis, i.e. a symmetric mode aptly called the "umbrella mode", then the stiffness is governed by the control servos, cables and bellcranks in the system joining the pilot's pedals to the tail rotor spider. However, if an asymmetric blade torsional behaviour occurs, then the loads to the servos, cables and bellcranks are reduced or completely cancelled and the stiffness of the spider and actuator shaft are the determining factors. For a pusher tail rotor such as the Westland Sea King, as the collective pitch on the tail rotor blades increases, the actuator shaft withdraws into the driveshaft and the effective stiffness will increase taking the torsional frequency with it. To complicate the story still further, at low pitch the blade lag stiffness will be very high since the blade will be bending in an essentially edgewise manner. The flapping frequency will be much lower, although still at a supercritical frequency, and any blade stiffness is in the flatwise sense and is therefore much lower than the edgewise stiffness in the lag direction. If the blades could rotate by a right angle, then the above arguments would apply but in the reverse order, i.e. the flapping frequency is now determined by the edgewise characteristics of the blade and the lag motion by the flatwise blade characteristics and what was the flapping hinge is now the lag hinge and vice versa. Distilling this argument means that as the collective pitch increases, the lag frequency falls and the flapping frequency increases. Figure 11.36, taken from Cook,[24] shows the fall in lag frequency of a typical tail rotor blade with collective pitch. Since the flapping frequency starts at about $1.05\Omega$, and then increases with collective pitch, a frequency coalescence at a high pitch angle can be seen to be a distinct possibility, leading to a potentially resonant condition.

We are therefore faced with the problems of two types of torsional motion each

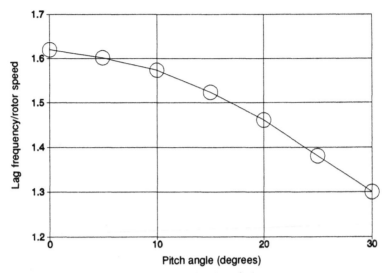

**Fig. 11.36.** Variation of lag frequency with collective pitch.

with its own frequency and together with flapping and lagging frequencies, and all three are varying with collective pitch angle. This multitude of variations has been analysed by Burton and Ellis[32] and Fig. 11.37 is based on their report of a rotor with a flap frequency of $1.04\Omega$ and a $\delta_3$ angle of $45°$. The figure plots the unstable regions when the lag and torsional frequencies are varied. There are two distinct regions labelled A and B.

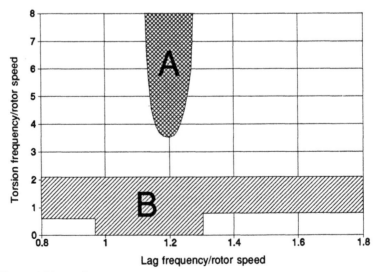

**Fig. 11.37.** Instability regions.

*Region A*. This is a mild instability brought about by the coalescence of the flapping and lagging frequencies with higher collective pitch angles. It is considered to be the so-called "buzz" behaviour and is primarily a flap and lag motion. It is placed at the higher torsional frequency (non-umbrella mode) and $\delta_3$ has the effect of raising the effective flapping frequency towards the lag frequency, triggering the instability.

*Region B*. This is a violent instability which is primarily a flutter-type motion consisting mainly of flap and torsion. It is affected by aft movement of the blade CG and an increase in $\delta_3$ angle. It occurs with the lower torsional frequency (umbrella mode). It is considered to be the "bang" instability observed on some aircraft which is characterised by a loud noise from the tail rotor and a sudden yawing motion in the direction of increasing tail rotor thrust. It usually causes severe damage to the rotor blades.

## Appendix: the intersection of a tail rotor blade with a main rotor vortex

The interaction of the tail rotor with the vortices left by the main rotor has already been highlighted with regard to handling and noise generation. The following analysis shows how the pattern of intersections can be calculated. It assumes an undistorted main rotor wake but illustrates how these effects can be examined and the important influence tail rotor rotation direction has on this phenomenon.

The geometry of the theoretical model is shown in Fig. 11.38. The origin is the centre of the tail rotor at zero time. The rotation of the tail rotor can be forward-at-the-top ($K_{ROT} = -1$) or aft-at-the-top ($K_{ROT} = +1$). The main rotor vortices are equispaced as the model assumes no movement relative to the air after shedding. The line of the main rotor vortices is displaced by $\bar{h}$ above the velocity vector of the tail rotor centre, and they are spaced at intervals of $\bar{s}$. The $X$ and $Y$ values are scaled to unit tail rotor radius.

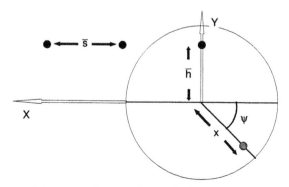

**Fig. 11.38.** Geometry of the vortex intersection analysis.

With this axis system, the position of a point of the tail rotor blade, having rotated by an angle of $\phi_{TR}$ (at radial station $x$), is given by

$$X = \mu\phi_{TR} - x\cos\psi$$
$$Y = -xK_{ROT}\sin\psi$$

(11.26)

Note that the application of the $K_{ROT}$ factor is applied to the rotation angle $\psi$, but as it is $\pm 1$ it has been extracted outside the trigonometric terms.

The azimuth position of a general blade is

$$\psi = \psi_B + \phi_{TR}$$

(11.27)

The tail rotor blades are sequenced in the direction of rotation by means of the azimuth angle $\psi_B$, which is zero on the reference blade, which is itself at $\psi = 0$ at the start of the time history.

The $N$th main rotor vortex is at

$$X_V = (N-1)\bar{s}$$
$$Y_V = \bar{h}$$

(11.28)

For an intersection the following must apply

$$X = X_V$$
$$Y = Y_V$$

(11.29)

i.e.

$$x\cos\psi = \mu\phi_{TR} - X_V$$
$$xK_{ROT}\sin\psi = -Y_V$$

(11.30)

Dividing the equations removes the $x$ term leaving the tail rotor rotation angle $\phi_{TR}$ to be determined. The algorithm used can show odd behaviour if the division is not expressed appropriately for the condition where an intersection is possible.

For the case of $\bar{h} < 0.5$

$$K_{ROT}\tan(\psi_B + \phi_{TR}) + \frac{Y_V}{\mu\phi_{TR} - X_V} = 0$$

(11.31)

Otherwise

$$K_{ROT}\cot(\psi_B + \phi_{TR}) + \frac{\mu\phi_{TR} - X_V}{Y_V} = 0$$

(11.32)

or $\bar{h} < 0.5$, the intersection will be along the central region of the tail rotor disc hen $\tan(\psi_B + \phi_{TR})$ is near to zero and (11.31) is well behaved. Otherwise the ıtersections will be towards the top or bottom of the disc when $\tan(\psi_B + \phi_{TR})$ can ɛcome very large and the problem ill-conditioned. Use of (11.32) avoids this ·oblem.

From the value of $\phi_{TR}$ the radial station $x$ can be determined

$$x = -\frac{K_{ROT} Y_V}{\sin(\psi_B + \phi_{TR})}, \quad \text{if } \sin(\psi_B + \phi_{TR}) \neq 0$$

$$x = \frac{\mu \phi_{TR} - X_V}{\cos(\psi_B + \phi_{TR})}, \quad \text{if } \cos(\psi_B + \phi_{TR}) \neq 0$$
(11.33)

he value of $x$ should lie in the range [0, 1]. Because of possible numerical prob ms, if a possible solution is located, the algorithm should perform a final check ı the distance between the blade position $(X, Y)$ and the vortex $(X_V, Y_V)$.

For efficiency in the calculation the search range can be restricted as shown in ıg. 11.39, where the values of $\phi_{TR}$ to be searched are given by

$$\phi_{TR\,min} = \frac{X_V - 1}{\mu}, \quad \phi_{TR\,max} = \frac{X_V + 1}{\mu}$$
(11.34)

The inter-vortex spacing term $\bar{s}$ is determined by the following. In a single ain rotor rotation, $N$ vortices are shed in a time of

$$\frac{2\pi}{\Omega_{Main}}$$
(11.35)

ı this time, the tail rotor has moved a distance of

$$\mu_{Tail}(\Omega_{Tail} R_{Tail}) \times \frac{2\pi}{\Omega_{Main}}$$
(11.36)

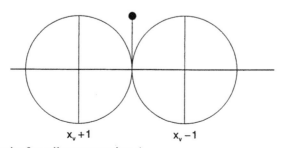

**g. 11.39.** Search limits for tail rotor rotation $\phi$.

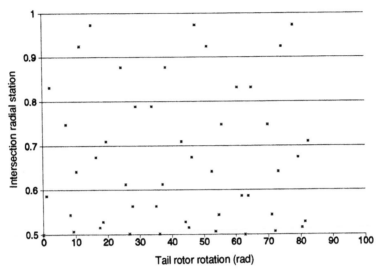

**Fig. 11.40.** Vortex intersections forward-at-the-top.

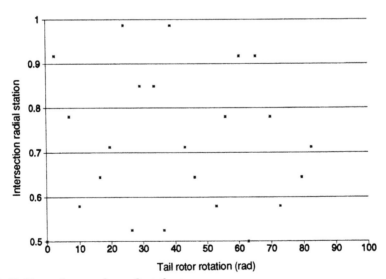

**Fig. 11.41.** Vortex intersections aft-at-the-top.

Therefore the vortex spacing $\bar{s}$ is given by

$$\bar{s} = \frac{1}{N}\left[\mu_{\text{Tail}}\,\frac{\Omega_{\text{Tail}}}{\Omega_{\text{Main}}} \times 2\pi\right] \qquad (11.37)$$

As an example of the method the following two cases have been run for the following data

$$\text{number of main rotor blades} = 4$$
$$\text{number of tail rotor blades} = 4$$
$$\mu = 0.3$$
$$\text{vortex line offset } \bar{h} = 0.5$$
$$\Omega_{\text{Main}} = 35 \text{ rad/s}$$
$$\Omega_{\text{Tail}} = 200 \text{ rad/s}$$

For a rotation of forwards-at-the-top ($K_{\text{ROT}} = -1$), see Fig. 11.40 and for aft-at-the-top ($K_{\text{ROT}} = +1$), see Fig. 11.41. The number of intersections is much reduced with aft-at-the-top rotation, hence the noise generation should be similarly reduced.

# 12

# Ground resonance

## Introduction

The helicopter main rotor must allow the rotor blades to lead and lag in the plane of the disc. This is necessary to alleviate the rotor hub components of Coriolis forcing generated by the blade flapping motion. The problem to be addressed in this chapter is that the existence of blade lag motion heralds the possibility of a resonant condition. If the blades lag in unison, the CG of the rotor head remains at the rotor centre, however a non-uniformity of the blade lag motion will cause the rotor CG to move off the hub centre and a type of stirring motion will be set up. The helicopter undercarriage contains springs and dampers and the airframe will therefore have natural frequencies of its own governed by the dynamic characteristics of the undercarriage (including tyres) and the mass and inertias of the fuselage. Consequently, a non-uniform lag motion of the blades will cause the movement of the main rotor CG off the axis of rotation and subject the fuselage and undercarriage to an oscillating force. Should this oscillating force be resolved from the rotor into the non-rotating plane of the fuselage, and the resulting frequency be on or close to that of the fuselage and undercarriage, then a resonant condition is a dangerous possibility. The offset CG will trigger a fuselage motion of rocking on its undercarriage. This will induce a lateral motion of the main rotor hub which in turn can then aggravate the blade lagging motion. In this way a divergent situation can arise with the inevitable result of severe damage to, or destruction of the aircraft with the obvious hazards to the flight crew. As will be shown later this situation, known as ground resonance, requires damping to be installed to the main rotor head/blade connection *and* the undercarriage. Wear and tear, maladjustment, or failure of any of these components can result in the occurrence of ground resonance, a typical result of which is shown in Plate 12.1.

The analysis of the ground resonance phenomenon begins with a simple model. It highlights the basic mechanism and the calculation of the frequency regions which are likely to cause problems. It can do no more than this, and in order to calculate the amounts of damping required to suppress any instability, a more detailed method of ground resonance is described.

The chapter continues with a discussion of the potentially worst fuselage/undercarriage condition for ground resonance and the calculation method to assess how near a given fuselage design is to its theoretically worst condition. The allied phenomenon of air resonance is introduced and the chapter concludes with a description of two methods of dealing with a potential ground resonance problem when running a prototype, or development helicopter rotor for the first time.

**Plate 12.1.** Westland Wessex after suffering ground resonance. (© British Crown Copyright 1993/MOD reproduced with the permission of the Controller of Her Britannic Majesty's Stationery Office.)

# Derivation of a simple ground resonance model

This method allows the offset rotor CG to be determined from the blade lagging motion and shows how the resulting frequencies can couple with those of the fuselage sitting on its undercarriage.

Each rotor blade is modelled by a concentrated mass connected by a light rod to the lagging hinge, as shown in Fig. 12.1. The location of the $k$th mass is given by

$$X_k = e_L R \cos\psi_k + r_g \cos(\psi_k + \zeta_k)$$
$$Y_k = e_L R \sin\psi_k + r_g \sin(\psi_k + \zeta_k)$$

(12.1)

For small $\zeta_k$ we can use the approximations $\cos\zeta_k \approx 1$ and $\sin\zeta_k \approx \zeta_k$ from which

$$X_k = e_L R \cos\psi_k + r_g \cos\psi_k - r_g \sin\psi_k \zeta_k$$
$$= (e_L R + r_g)\cos\psi_k - r_g\zeta_k \sin\psi_k$$
$$Y_k = e_L R \sin\psi_k + r_g \sin\psi_k + r_g \cos\psi_k \zeta_k$$
$$= (e_L R + r_g)\sin\psi_k + r_g\zeta_k \cos\psi_k$$

(12.2)

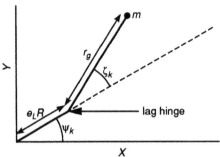

**Fig. 12.1.** Coordinates of general blade.

For a rotor with $N$ such blades, if the rotor CG is located at $(X_c, Y_c)$ then we have

$$X_c = \frac{1}{mN} \sum_{k=1}^{N} mX_k, \qquad Y_c = \frac{1}{mN} \sum_{k=1}^{N} mY_k \qquad (12.3)$$

A blade is selected to be at an azimuth of $\psi$ and so the $k$th blade is at an azimuth angle of

$$\psi_k = \psi + (k-1)\frac{2\pi}{N} \qquad (12.4)$$

where the inter-blade spacing is $\phi = 2\pi / N$.

Then from Eqns (12.2) and (12.3) we find the following for the $X$ component of the rotor CG

$$X_c = \frac{1}{mN} \sum_k m[(e_L R + r_g)\cos\psi_k - r_g \zeta_k \sin\psi_k]$$

$$= \frac{1}{N}\left[(e_L R + r_g)\sum_k \cos\psi_k - r_g \sum_k \zeta_k \sin\psi_k\right] \qquad (12.5)$$

$$= -\frac{r_g}{N}\sum_k \zeta_k \sin\psi_k$$

where the following result has been used

$$\sum_k \cos\psi_k = 0 \qquad (12.6)$$

Similarly, for the $Y$ component of the rotor CG, we obtain

$$Y_c = \frac{r_g}{N} \sum_k \zeta_k \cos\psi_k \qquad (12.7)$$

If we now assume the lagging behaviour of the blades to be a simple harmonic motion (SHM) of amplitude $\zeta_0$ and frequency $K\Omega$, we have for the lag angle of the blade at an azimuth angle of $\psi_k$

$$\zeta_k = \zeta_0 \cos(K\psi_k) \qquad (12.8)$$

Substituting this expression into Eqns (12.5) and (12.7) we find

$$\begin{aligned} X_c &= -\frac{r_g\zeta_0}{N} \sum_k \cos K\psi_k \cdot \sin\psi_k \\ &= -\frac{r_g\zeta_0}{2N} \sum_k [\sin(K+1)\psi_k - \sin(K-1)\psi_k] \\ Y_c &= \frac{r_g\zeta_0}{N} \sum_k \cos K\psi_k \cdot \cos\psi_k \\ &= \frac{r_g\zeta_0}{2N} \sum_k [\cos(K+1)\psi_k + \cos(K-1)\psi_k] \end{aligned} \qquad (12.9)$$

We now require summations of the form

$$\sum_{k=1}^{N} \cos n\psi_k \qquad (12.10)$$

where $n$, in this instance, is not necessarily an integer.
To perform this sum we define the following summations

$$\begin{aligned} C &= \sum_{k=1}^{N} \cos n(\psi + [k-1]\phi) \\ S &= \sum_{k=1}^{N} \sin n(\psi + [k-1]\phi) \end{aligned} \qquad (12.11)$$

These can be combined into a single complex quantity

$$\begin{aligned} C + iS &= \sum_{k=1}^{N} (\cos + i\sin)\{n(\psi + [k-1]\phi)\} \\ &= \sum_{k=1}^{N} e^{in(\psi + [k-1]\phi)} \end{aligned} \qquad (12.12)$$

This is a geometric progression of $N$ terms, with the first term $e^{in\psi}$, and the multiplying factor $e^{in\phi}$. This results in

$$C + iS = e^{in\psi}\left(\frac{e^{iN\phi n} - 1}{e^{i\phi n} - 1}\right) \tag{12.13}$$

Noting that

$$e^{iN\phi n} - 1 = e^{iN\phi n/2}\left(e^{iN\phi n/2} - e^{-iN\phi n/2}\right)$$

$$= e^{iN\phi n/2} 2i\sin\left(\frac{N\phi n}{2}\right)$$

$$e^{i\phi n} - 1 = e^{i\phi n/2}\left(e^{i\phi n/2} - e^{-i\phi n/2}\right) \tag{12.14}$$

$$= e^{i\phi n/2} 2i\sin\left(\frac{\phi n}{2}\right)$$

From which we obtain

$$C + iS = \exp\left[in\left(\psi + \frac{N-1}{2}\phi\right)\right] \cdot \frac{2i\sin(nN\phi/2)}{2i\sin(n\phi/2)}$$

$$= \frac{\sin(nN\phi/2)}{\sin(n\phi/2)} \cdot (\cos + i\sin)\left[n\left(\psi + \frac{(N-1)\phi}{2}\right)\right] \tag{12.15}$$

Since the inter-blade spacing, $\phi$, is given by

$$\phi = \frac{2\pi}{N} \tag{12.16}$$

then on substitution in (12.15)

$$C + iS = \frac{\sin(n\pi)}{\sin(n\pi/N)} \cdot (\cos + i\sin)\left[n\left(\psi + \frac{(N-1)\pi}{N}\right)\right] \tag{12.17}$$

From which we obtain on equating real and imaginary parts

$$C = \frac{\sin(n\pi)}{\sin(n\pi/N)} \cdot \cos\left[n\left(\psi + \frac{(N-1)\pi}{N}\right)\right]$$

$$S = \frac{\sin(n\pi)}{\sin(n\pi/N)} \cdot \sin\left[n\left(\psi + \frac{(N-1)\pi}{N}\right)\right] \tag{12.18}$$

For brevity, we define

$$S_{K+1} = \frac{\sin(\overline{K+1}\pi)}{\sin\left(\overline{K+1}\dfrac{\pi}{N}\right)}$$

$$S_{K-1} = \frac{\sin(\overline{K-1}\pi)}{\sin\left(\overline{K-1}\dfrac{\pi}{N}\right)} \qquad (12.19)$$

$$\Phi = \frac{N-1}{N}\pi$$

then

$$X_c = -\frac{r_g \zeta_0}{2N}[S_{K+1}\sin(K+1)(\psi+\Phi) - S_{K-1}\sin(K-1)(\psi+\Phi)]$$
$$\qquad (12.20)$$
$$Y_c = \frac{r_g \zeta_0}{2N}[S_{K+1}\cos(K+1)(\psi+\Phi) + S_{K-1}\cos(K-1)(\psi+\Phi)]$$

This means that the centre of mass of the $N$ blades can be represented by two masses rotating around the shaft at different radii and rotational speeds

$$\text{mass} = mN$$

$$\text{radius} = \frac{r_g \zeta_0}{2N} \cdot S_{K \pm 1} \qquad (12.21)$$

$$\text{rotational speed} = (K \pm 1)\Omega$$

A typical value for $K$ is 0.3 from which $K+1 = 1.3$ and $K-1 = -0.7$, the former result is known as the progressive mode and it rotates with the rotor direction. The latter is the regressive mode and rotates against the rotor direction.

With reference to Figs 12.2 and 12.3 the variation of the progressive and regressive mode frequencies are plotted against rotor speed. In addition, in Fig. 12.3 the fuselage frequency is plotted for two types of frequency crossings namely, $A$ and $C_2$ or $B$ and $C_1$.

A more detailed analysis shows that the intersections $A$ and $B$ are *stable*, whilst $C_1$ and $C_2$ are *unstable* and are therefore the source of ground resonance.

# Advanced ground resonance model

The simple analysis just described shows the manner in which the rotor blade lagging and the fuselage oscillating on its undercarriage cause resonance. However, it does not provide any means of assessing the severity of the instability, and

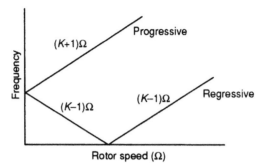

**Fig. 12.2.** Progressive and regressive frequencies.

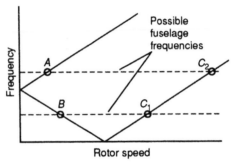

**Fig. 12.3.** Coupling of undercarriage frequency.

the degree of damping required to suppress it. A more realistic theoretical model is required and is shown in Fig. 12.4. Here the rotor hub is allowed one degree of translational freedom and viscous (linear) damping can be included both on the fuselage translation and the blade rotation about the lagging hinge.

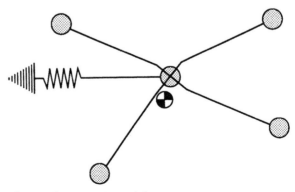

**Fig. 12.4.** Advanced ground resonance model.

Again each of the $N$ blades is modelled as a lumped mass $m$. As the rotor head motion is now included, its mass must be included. This is specified as $M$. The blade natural lag frequency is given by $\omega_\zeta$, and the fuselage natural frequency by $\omega_{FUS}$.

In the previous analysis, the lag frequency was expressed as a multiple $(K)$ of the rotor speed $\Omega$. This is reasonable for an articulated rotor, but a semi rigid rotor, having flexible elements, will not have this simple linear variation of lagging frequency with rotor speed. To avoid any problems of confusion, the lagging frequency of $K\Omega$ is now replaced by $\omega_\zeta$ then $(K\pm1)\Omega$ becomes $\omega_\zeta\pm\Omega$.

The solution of the more advanced method is not described in detail here, but the equations of motion are set up using the Gibbs–Appell method, described by Pars.[33] There are $N+1$ equations of motion corresponding to the $N$ blades and the single degree of fuselage freedom. Unfortunately, the equations are non-linear with periodic coefficients, however Coleman and Feingold[34] derived a transformation to the so-called Coleman coordinates which removes this periodicity of the coefficients. The $N+1$ equations reduce to three linear, second-order differential equations of motion. The usual method of substituting a SHM variation on all the three unknown variables at a given frequency results in a sextic polynomial whose solution gives the frequencies and damping (negative or positive) of the system. As the sextic polynomial has real coefficients, complex solutions occur in complex conjugate pairs. If the polynomial is solved for the six roots, over a range of rotor speeds, some conditions cause only imaginary solutions which describe steady-state SHM vibration. However, a range of rotor speeds can occur whereby two solutions have real parts, one being positive. This corresponds to an instability and is therefore the origin of ground resonance. To show this, a semi-rigid rotor helicopter, based on the Westland Lynx, is analysed by this method and the results are presented.

The input data are

$$\text{static lag frequency} = 15.22 \text{ rad/s}$$
$$\text{static fuselage frequency} = 12.0 \text{ rad/s}$$
$$\text{lag hinge offset} = 1.22 \text{ m}$$
$$\text{rotor radius} = 6.4 \text{ m}$$
$$\text{fuselage effective mass} = 500 \text{ kg}$$
$$\text{blade mass} = 24.8 \text{ kg}$$
$$\text{number of blades} = 4$$

Figure 12.5 shows the imaginary (frequency) parts of the solutions against rotor speed. Figure 12.6 shows the corresponding positive real parts (stability/instability) also against rotor speed. In the figures a positive value corresponds to an unstable root, whilst a negative value indicates a stable root. The region of instability is clearly seen accompanied by a coalescence of the resonant frequencies. The maximum instability can be monitored, to see the degree of instability and the amount of damping which needs to be fitted to the rotor hub to ensure stability at all rotor speeds. The manner in which the fuselage reacts is embodied in the effective mass term $M$. Variation of this affects the instability (real part) as can be seen

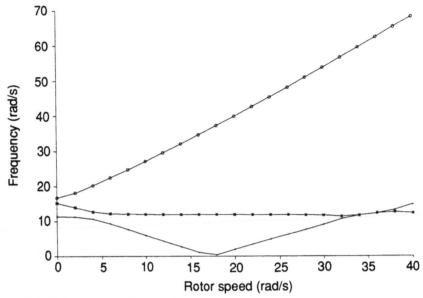

**Fig. 12.5.** Coleman mode frequencies.

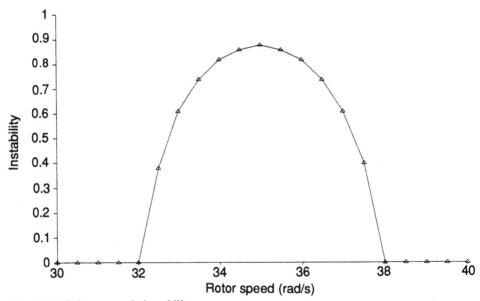

**Fig. 12.6.** Coleman mode instability.

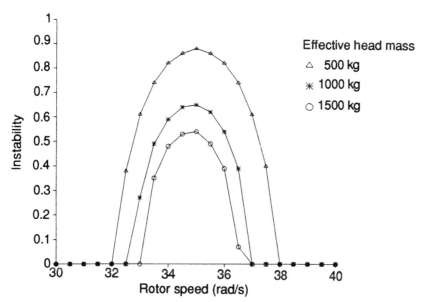

**Fig. 12.7.** Coleman mode instabilities—effect of head mass.

in Fig. 12.7. Three values of the effective head mass have been analysed and the degree of instability can be seen to increase with decreasing head mass. In other words, the lighter an airframe, the more prone the aircraft is to ground resonance.

## Suppression of ground resonance

The method has been used on the same aircraft data to calculate the amount of damping required to suppress the ground resonance instability over the complete rotor speed range. Figure 12.8 shows the instability variation for different amounts of damping, both on the rotor head (lag dampers) and the fuselage (undercarriage oleos and tyres). It can be seen that placing increasing amounts of damping on either the rotor head/blades or the fuselage in isolation is not at all effective. However, the cases where the damping is applied to *both* rotor head and fuselage are seen to contain the instability, and ultimately to completely suppress it. This conclusion is derived by Coleman and Feingold.[34] It is expressed, very importantly, as a minimum value for the product of the damping values, i.e.

$$\text{damping}_{\text{fuselage}} \cdot \text{damping}_{\zeta} > \text{minimum value} \qquad (12.22)$$

The standard method is therefore the addition of damping; applying it to both the rotor head and the fuselage. Figure 12.9 shows the stability boundary illustrating this damping allocation. The lag frequency has an important influence on the boundary, see Fig. 12.10, so as it increases towards $1\Omega$, the required damping decreases.

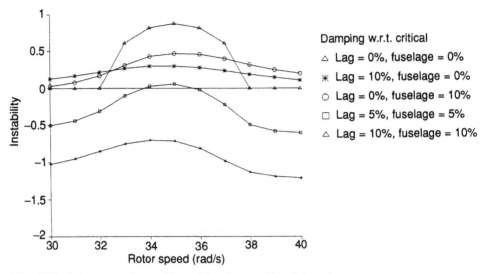

**Fig. 12.8.** Coleman modes—effect of fuselage and head damping.

If the lag frequency exceeds $1\Omega$ then the lag motion is termed "supercritical", when both the $\omega_\zeta \pm \Omega$ values are positive and there is therefore no regressing lag mode. Although an increase in lag frequency may therefore seem advantageous, a drawback exists if the lag frequency is not significantly different than $1\Omega$ ($1.05\Omega$ say). In this situation, the $1\Omega$ air and Coriolis loads will be amplified.

The analysis of ground resonance is made much more tractable with the use of a viscous damper in which the damper force is directly proportional to the damper velocity. In reality, dampers exhibit different, and more complicated, characteristics and to retain the benefits of a linear damping law, the concept of an equivalent viscous damper is used.

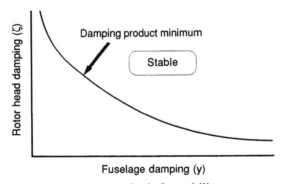

**Fig. 12.9.** Coleman modes—minimum criteria for stability.

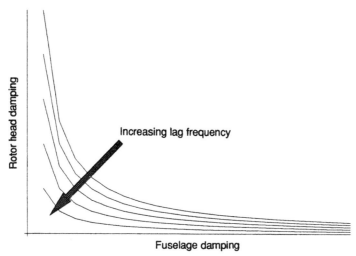

**Fig. 12.10.** Coleman modes—effect of lag frequency on required damping.

# Damping

## *Blade/head damping*

The simplest blade lag damper is of the frictional type, whereby the damping force is of constant value, and always opposing the motion. It suffers from a "sticking effect" as the motion changes direction, which can cause severe vibration problems with rotor engagement accelerations on the ground, and power changes in flight which cause a change in steady lag angle. An irony exists in that fitting a device to a helicopter to ensure stability on the ground, could cause an embarrassing decrease in the fatigue life of a blade during flight. A common design of a modern rotor blade lag damper consists of a parabolic variation of damper load with velocity up to a certain value followed by a constant value of force thereafter. This provides a characteristic not unlike a friction damper in the main, without the sticking problems.

For both types of damper, the effectiveness of the damping decreases at higher forcings, so a stable condition could become unstable if the blade motion is allowed to exceed a given value. Because of this, accurate estimates of the values of hub acceleration occurring in service life will be necessary for the designer to assess whether the blade swing angles will be dangerously high. Figures 12.11 and 12.12 show the effect of forcing amplitude on the effective linear (viscous damping) of frictional and hydraulic dampers. The effective damper is a viscous damper of the appropriate characteristic to absorb the same energy per cycle. The variation of its effectiveness with increasing amplitude is seen in the figures for a Coulomb (or friction) damper and a typical hydraulic type. The fall off at the higher blade swing angles can be seen giving rise to the hazardous situation of an unstable behaviour being aggravated by a loss of effective damping.

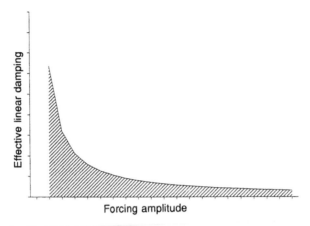

**Fig. 12.11.** Effective linear Coulomb damper.

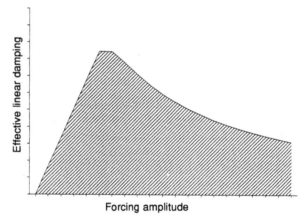

**Fig. 12.12.** Effective linear hydraulic damper.

## *Fuselage damping*

At the project stage, little is known about the undercarriage characteristics (namely, damping and stiffness). One way to avoid this problem is to assume the worst case which corresponds to

1.  The frequency coalescence between blade lag and fuselage motions occurs at maximum operating rotor speed. (It can be seen in Fig. 12.13 that the instability is worse as the fuselage frequency increases, providing the frequency coalescence does not occur above the maximum rotor speed.)
2.  Ratio of head mass to fuselage mass affects the damping required in a direct manner. (So a heavy rotor system and a light fuselage is worst.) In reality, fuselage motion will consist of a mixture of surge/sway translation and pitch/

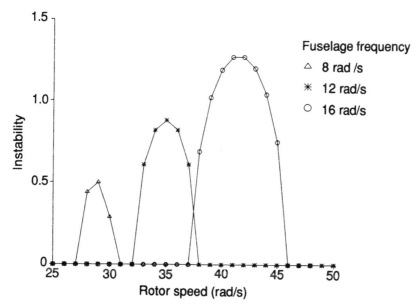

**Fig. 12.13.** Effect of fuselage frequency on Coleman mode instability.

roll rotation and so these degrees of freedom will need to be included in the analysis.

3.  When analysing the helicopter undercarriage characteristics the fuselage modes must be considered. This will include the effects of the fuselage inertias and flexibilities. The calculations will search for the dynamic behaviour of the fuselage and undercarriage combination which possesses significant rotor head motion in the rotor plane. (Effective head mass.)

4.  The dynamic characteristics of the undercarriage can change with load (oleos, tyres, etc.) so a demonstration of freedom from ground resonance must include the full available range of rotor thrust and therefore loads in the undercarriage legs. The ground surface condition is also important as this will determine the frictional loads applied to the tyres.

Oleos have significant non-linear stiffness characteristics and it is possible that the oleo can stick in a high lift condition where the load in the oleo is below a preset break out load and the only damping will come from the tyres. (This in turn necessitates correct inflation.) A "balance spring" can be included in the oleo design to allow oleo movement for loads about any datum level.

## Effective head mass

In the analysis of the ground resonance model the fuselage has been modelled by a lumped mass at the rotor centre. This "effective mass" must be correctly sized to accurately reflect the motion of the fuselage. Considering its single degree of

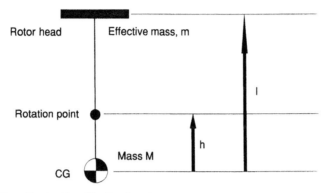

**Fig. 12.14.** Coordinates for effective head mass.

freedom, the hub motion will be determined by a translation and a rotation of the fuselage. If an energy approach such as Lagrange's equations are considered, then the kinetic energy of the fuselage motion must be included. This means that the effective mass must correctly recreate the kinetic energy of the fuselage.

Consider a fuselage whose motion is limited to roll and sway only as shown in Fig. 12.14. The fuselage is translating and rotating, and will have an effective rotation point as indicated in the figure. Using the following notation

$$\text{fuselage mass} = M$$
$$\text{fuselage moment of inertia about CG} = I$$
$$\text{height of effective rotation point above CG} = h$$
$$\text{height of rotor hub above CG} = l$$

For equal kinetic energy, since either system is oscillating at the same angular rate, the moments of inertia of the effective mass at the rotor hub location and the complete fuselage about the effective rotation point must be equal giving

$$m(h-l)^2 = Mh^2 + I \qquad (12.23)$$

i.e.

$$m = \frac{I + Mh^2}{(h-l)^2} \qquad (12.24)$$

The single variable is the effective rotation point height $h$, which can be shown to give a minimum value for $m_{\text{EFF}}$, when

$$h_{\text{min}} = -\frac{I}{Ml}, \qquad m_{\text{min}} = \frac{MI}{I + Ml^2} \qquad (12.25)$$

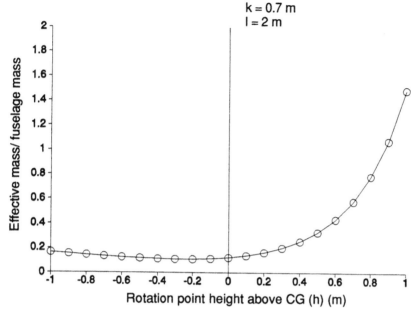

**Fig. 12.15.** Variation of effective head mass with rotation centre.

If $k$ is the radius of gyration of the fuselage about its CG then

$$I = Mk^2 \tag{12.26}$$

Equation (12.25) can be re-expressed as

$$h_{min} = -\frac{k^2}{l}, \qquad \frac{m_{min}}{M} = \frac{k^2}{k^2 + l^2} \tag{12.27}$$

Figure 12.15 shows the variation of effective mass with rotation point location ($h$) and the minimum can be clearly seen. By this simple analysis the potentially worst fuselage characteristics can then be evaluated.

## Fuselage modes

Usually, however, the fuselage design specifies an undercarriage geometry and oleo spring rates, in which case the actual fuselage modes of vibration can be found. From this the effective head mass for a given undercarriage specification can be calculated. Additionally, the minimum effective head mass can be calculated, and the proximity of a given fuselage design to the worst ground resonance condition can then be determined.

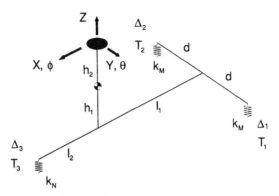

**Fig. 12.16.** Geometry for fuselage modes.

As an example of such a calculation, Fig. 12.16 shows a simple undercarriage/fuselage combination. A tricycle wheel arrangement is shown with two main wheels and a single nose wheel. The three oleos are numbered, and each oleo is considered as a linear spring acting in the vertical direction only. The data for each oleo are given in Table 12.1. The motion is limited to heave, roll and pitch about CG.

**Table 12.1.** Oleo data

| Oleo No. | Location | Extension | Spring rate | Load |
|----------|----------|-----------|-------------|------|
| 1 | Main port | $\Delta_1$ | $k_m$ | $T_1$ |
| 2 | Main starboard | $\Delta_2$ | $k_m$ | $T_2$ |
| 3 | Nose | $\Delta_3$ | $k_n$ | $T_3$ |

The geometry is specified by

$$\text{semi-wheel track} = d$$
$$\text{location of main gear aft of CG} = l_1$$
$$\text{location of nose gear forward of CG} = l_2$$
$$\text{height of CG above datum} = h_1$$
$$\text{height of rotor head above CG} = h_2$$

The fuselage kinematics, relative to the CG, are specified by

$$\text{mass} = M$$
$$\text{roll inertia} = I_R$$
$$\text{pitch inertia} = I_P$$

If the fuselage, relative to the CG, is displaced $Z$ vertically, $\phi$ in roll, and $\theta$ in pitch then we have for the oleo extensions

$$\Delta_1 = Z + d\phi + l_1\theta$$
$$\Delta_2 = Z - d\phi + l_1\theta \qquad (12.28)$$
$$\Delta_3 = Z - l_2\theta$$

The oleo tension loads are then

$$T_1 = k_M \Delta_1$$
$$T_2 = k_M \Delta_2 \qquad (12.29)$$
$$T_3 = k_N \Delta_3$$

The equations of motion can now be assembled.

Heave (Z)

$$MÏ = -k_M(\Delta_1 + \Delta_2) - k_N\Delta_3$$
$$= -k_M 2(Z + l_1\theta) - k_N(Z - l_2\theta) \qquad (12.30)$$

Roll ($\phi$)

$$I_R\ddot{\phi} = k_M(\Delta_2 - \Delta_1)d$$
$$= -2d^2 k_M \phi \qquad (12.31)$$

Pitch ($\theta$)

$$I_P\ddot{\theta} = -k_M(\Delta_1 + \Delta_2)l_1 + k_N\Delta_3 l_2$$
$$= -k_M l_1 2(Z + l_1\theta) + k_N l_2(Z - l_2\theta) \qquad (12.32)$$

Equations (12.30)–(12.32) can be cast into a matrix form.
Defining the following matrices

$$M_{FUS} = \begin{bmatrix} M & 0 & 0 \\ 0 & I_R & 0 \\ 0 & 0 & I_P \end{bmatrix}$$

$$K_{FUS} = -\begin{bmatrix} 2k_M + k_N & 0 & 2k_M l_1 - k_N l_2 \\ 0 & 2k_M d^2 & 0 \\ 2k_M l_1 - k_N l_2 & 0 & 2k_M l_1^2 + k_N l_2^2 \end{bmatrix} \qquad (12.33)$$

the equations of motion become

$$M_{\text{FUS}} \begin{bmatrix} \ddot{Z} \\ \ddot{\phi} \\ \ddot{\theta} \end{bmatrix} + K_{\text{FUS}} \begin{bmatrix} Z \\ \phi \\ \theta \end{bmatrix} = 0 \qquad (12.34)$$

We seek an SHM solution of the form

$$\begin{bmatrix} Z \\ \phi \\ \theta \end{bmatrix} = \begin{bmatrix} a_Z \\ a_\phi \\ a_\theta \end{bmatrix} e^{i\omega t} \qquad (12.35)$$

where $\omega$ is the modal frequency and $a_Z$, $a_\phi$ and $a_\theta$ are the relative displacement amplitudes of the three variables.

Substituting (12.35) into (12.34) gives the following result

$$[-\omega^2 M_{\text{FUS}} + K_{\text{FUS}}] \begin{bmatrix} a_Z \\ a_\phi \\ a_\theta \end{bmatrix} = 0 \qquad (12.36)$$

Since $M_{\text{FUS}}$ is non-singular this can be recast as

$$[M_{\text{FUS}}^{-1} K_{\text{FUS}} - \omega^2 I] \begin{bmatrix} a_Z \\ a_\phi \\ a_\theta \end{bmatrix} = 0 \qquad (12.37)$$

This result gives the square of the modal frequencies as the eigenvalues of the matrix $[M_{\text{FUS}}^{-1} K_{\text{FUS}}]$ with the mode shapes as the corresponding eigenvectors. As $M_{\text{FUS}}$ is diagonal its inversion is trivial, and since $K_{\text{FUS}}$ is symmetric the matrix $[M_{\text{FUS}}^{-1} K_{\text{FUS}}]$ is real and symmetric making the calculation of the eigenvalues and eigenvectors straightforward.

If we define an eigenvector $m_d$ as

$$m_d = \begin{bmatrix} a_Z \\ a_\phi \\ a_\theta \end{bmatrix} \qquad (12.38)$$

The velocity of the rotor head will be given by

$$\omega \begin{bmatrix} h_2 a_\theta \\ -h_2 a_\phi \\ a_z \end{bmatrix} = \omega M_D m_d \qquad (12.39)$$

where

$$M_D = \begin{bmatrix} 0 & 0 & h_2 \\ 0 & -h_2 & 0 \\ 1 & 0 & 0 \end{bmatrix} \qquad (12.40)$$

The kinetic energy of the effective head mass is therefore given by

$$\begin{aligned} \tfrac{1}{2} m_{\text{EFF}} \omega^2 m_d^T M_D^T M_D m_d &= \tfrac{1}{2} M (a_z \omega)^2 + \tfrac{1}{2} I_R (a_\phi \omega)^2 + \tfrac{1}{2} I_P (a_\theta \omega)^2 \\ &= \tfrac{1}{2} m_d^T M_{\text{FUS}} m_d \omega^2 \end{aligned} \qquad (12.41)$$

The effective mass $m_{\text{EFF}}$ can now be determined for each modal frequency and shape via

$$m_{\text{EFF}} = \frac{m_d^T M_{\text{FUS}} m_d}{m_d^T M_D^T M_D m_d} \qquad (12.42)$$

and the proximity of each mode to the worst condition can be assessed. Additionally the benefits of any modifications to the undercarriage can be assessed very readily.

## Air resonance

Ground resonance is the result of the proximity of a fuselage natural frequency to that of the regressing rotor blade lag mode. In flight it is difficult to envisage how a phenomenon such as this could occur as the fuselage will apparently have no restraint such as that provided by its undercarriage. In fact, the equivalent air resonance phenomenon can exist[17] and the essential reasons are that a regressing cyclic flap mode can be generated which couples with the airframe roll and pitch motion to cause the fuselage to execute pendulum type, or slow gyroscopic modes, at frequencies close to the regressing lag mode. In other words, the blade cyclic flapping motion couples with the fuselage motion to give an oscillatory vibration of the head in the plane of the rotor. (This is what the undercarriage provides in ground resonance.) This motion now couples with the blade lagging motion to trigger a resonance. Air resonance can occur with a semi-rigid rotor but an articulated rotor cannot produce the fuselage pendulum modes with appropriate frequencies. In this case, the frequencies are so low as to prevent a coalescence with the regressing lag mode frequency at the rotor speeds normally encountered in flight.

The "stiffness" of the fuselage is based on the rotor flapping, which is heavily damped via the aerodynamics. However, it has already been stated from Coleman and Feingold[34] that the product of the rotor head and fuselage dampings is the most important factor, rather than individual values. Indeed, high damping in the fuselage and low damping in the rotor head, causes the unstable region to widen. Hence, head damping is still necessary but can be quite small. The situation of high fuselage stiffness and low head damping is one where air resonance could be a problem which unfortunately is typical of a semi-rigid rotor design, including hingeless and bearingless rotors.

In air resonance, the coupling between blade flapping and lagging motions caused by Coriolis forces are now of importance and such coupling of the two motions can in certain instances aid stability.

The amount of damping required to control air resonance is typically much less than that required for ground resonance. The danger is for a ground resonance solution to be adopted which does not result in additional lag dampers being fitted but instead uses the fuselage/undercarriage characteristics to avoid a frequency coalescence. Here, air resonance is a real possibility.

## Avoidance of ground resonance

While the results of appropriate ground resonance calculations can justify a given helicopter design, full verification on an aircraft is necessary before flight can be attempted. A full ground run test sequence must be carried out, which for reasons already explained before will be conducted at the minimum effective head mass to give the greatest potential for problems. This will normally be at a light airframe weight, however the situation can arise whereby at such a light aircraft weight the undercarriage will behave differently to a normal loading value, and an unrepresentative situation will occur. For instance, helicopter undercarriages can have preloaded components giving large changes in stiffness and damping as the oleo loads increase. It is naturally imperative that ground resonance must be avoided during prototype testing of an airframe. Whilst every human effort can be directed at avoiding ground resonance, an escape/checking mechanism is invariably used for testing the safety of the aircraft. Two types of procedure are detailed below.

### Snatch rig

*Principle.* With this system the fuselage is restrained by cables joining gearbox lugs and other strengthened members to hard points on the ground via hydraulic jacks (see Fig. 12.17). The aim is to artificially raise the fuselage frequency clear of the regressive lag frequency, if an instability is sensed. Figure 12.18 shows how, by raising the fuselage frequency sufficiently, the resonant condition can be moved above the rotor speed maximum.

Possible fuselage restraints

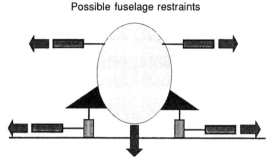

**Fig. 12.17.** Typical snatch ring configuration.

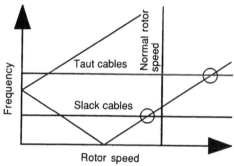

**Fig. 12.18.** Principle of the snatch rig.

*Method in practice.* The fuselage frequencies are first checked and if they are as predicted the aircraft can be moved to the hard standing used for the ground trials. The fuselage is then connected to the ground by cables which are kept initially slack. The rotor is engaged and the run up proceeds in steps. At each rotor speed the pilot will be providing small inputs by the cyclic stick to the rotor to see if an instability can be induced. This will continue until the maximum rotor speed is attained. This testing sequence is then repeated with steadily increasing main rotor thrust values until the helicopter is at the point of take-off. Should the pilot or ground test crew sense a problem, a "panic button" is pressed which powers the jacks to pull the cables taught, which moves the fuselage frequencies clear of a potential problem area; ideally it is never used but merely serves as a precaution.

## Moving block

*Principle.* This method[35] provides a means of assessing the damping in a system by analysing the response to an input. It concentrates on a particular frequency which is determined by a spectral analysis of the response. The moving block method was originally developed by the Lockheed–California Company in the early 1970s for use in rotorcraft stability investigation but can also be used for flutter testing. For a helicopter, an input is readily applied by a cyclic stick stir, which via the flapping motion caused by the cyclic stir and the Coriolis coupling between flap and lag motion will trigger lead/lag motion of the rotor blades.

*Description of the method.* The analytical basis of the moving block analysis is based on the transient response of a single mode

$$f(t) = Ae^{\sigma t} \sin(\omega t + \phi) \tag{12.43}$$

where $A$ is a constant, $\sigma$ is the exponential decay/growth, $\omega$ is the circular frequency and $\phi$ is a phase term.

A Fourier transform is performed on (12.43) over a finite time interval of $T$

$$F(\omega,\ \tau) = \int_{\tau}^{\tau+T} Ae^{\sigma t} \sin(\omega t + \phi)e^{-i\omega t}\ dt \tag{12.44}$$

This is equivalent to a full Fourier transform of (12.43) with a "boxcar" function applied

$$B(t) = \begin{pmatrix} 1, & \tau \leqslant t \leqslant \tau + T \\ 0, & \text{otherwise} \end{pmatrix} \tag{12.45}$$

$F(\omega,\ \tau)$ is referred to as the *moving block function.*

For small damping $(\zeta)$ the natural logarithm of the moving block function is given by

$$\ln|F(\omega,\ \tau)| = -\zeta\omega\tau + \tfrac{1}{2}\sin 2(\omega\tau + \phi) + \text{constant} \tag{12.46}$$

thus, the function $\ln|F|$ is a combination of a straight line of slope $\sigma = -\zeta\omega$ with an additional oscillatory component of twice the analysing frequency superimposed. From this the value of the damping $\zeta$ can be determined. The method is shown diagrammatically in Fig. 12.19.

*Method in practice.*
1. The response to an input (cyclic stick stir) is recorded.
2. A full Fourier transform (FFT) is performed on the signal to determine the frequency of interest to be used in the analysis.
3. The block length $(T)$ is chosen, and is much less than the signal duration. This will correspond to $N_b$ samples.
4. The left-hand side of (12.46) is calculated for $\tau = 0$.
5. This calculation is repeated for a range of values of $\tau$ by stepping along the samples one at a time. (This is the moving block.) For example, if $\Delta t$ is the period between samples, then $\tau = n\Delta t$, where $n = 0, 1, 2, 3, \ldots, N - N_b$ and $N$ is the total number of samples in the signal.
6. A least squares linear regression analysis is performed on $\ln|F(\omega,\ \tau)|$ vs $\tau$.
7. The slope of the straight regression line is determined, hence giving the system damping.

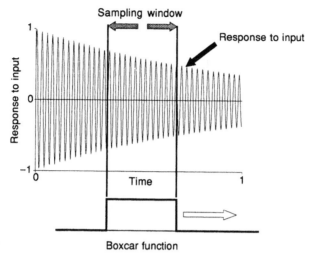

**Fig. 12.19.** Principle of the moving block analysis.

*Application to ground resonance.* The ground resonance investigation consists of a slow progressive increase of the rotor speed. At each value, a cyclic stick stir is initiated and the resulting response (lag angle or fuselage vibration) is recorded. From this signal the damping can be determined by the method. The damping can then be plotted against rotor speed allowing the overall behaviour of the aircraft to be closely monitored and a potential instability ($\zeta = 0$) to be predicted, in advance. As with the snatch rig, this can be repeated for a progressively increasing main rotor thrust until the helicopter is at the point of lifting off the ground. In this way the aircraft can be cleared. Note that the aircraft conditions used for the test will usually be set to make it the worst possible for an initial check.

# Rotor speed governing

## Requirement for rotor speed governing

The pilot's controls on early helicopters were main rotor collective and cyclic pitch, tail rotor collective pitch and engine throttle. The throttle was usually a twist grip control fitted to the main rotor collective pitch lever. The control of the engine via a throttle places an extra item on the pilot's workload and any device which relieves the pilot of this responsibility must be seen as an advantage. The engine control is required for the single purpose of supplying the correct amount of power to the rotor according to a specified law and so some type of governor can be contemplated.

The mechanism which the governor can use as a datum is the normal operating rotor speed and there are definite advantages in keeping the rotor speed within close limits. The main rotor is usually the greatest vibration source in the helicopter and the dynamic characteristics of both rotor and fuselage must be kept within a tight tolerance frequency band if resonances or amplified responses within the helicopter are to be avoided. Since the rotor blade natural frequencies are, in the main, dependent on rotor speed, the designer of the helicopter must assume that the forcing frequencies, and hence the rotor speed, are precisely controlled.

Vibration absorbers and isolators are important devices for improving the flight characteristics of a helicopter, however, their tuning is critical. Bifilar/pendulum rotor head vibration absorbers adjust their operating frequency in sympathy with the rotor speed. Conversely, fixed frequency vibration absorbers cannot retune and so their performance is very dependent on a closely monitored main rotor speed.

## Mechanism of control

The control of the engine is normally by means of a fuel valve, therefore the governor must sense a change in rotor speed brought about by a change in torque demand and adjust the fuel valve accordingly. This will continue until the balance of engine torque and rotor torque demand occurs. As the governor will only operate when the rotor speed is displaced from a preset datum value, the new steady-state will be at a different rotor speed than the original value. This change in rotor speed from the datum value is known as the *static droop*. The governor characteristics can be plotted as the torque demanded from the engines by the governor against rotor speed, and a typical example is shown in Fig. 13.1.

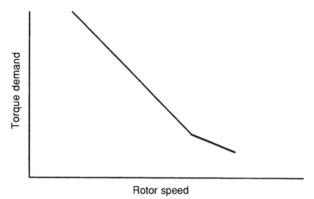

**Fig. 13.1.** Basic droop law.

There is a change in the slope at low engine/rotor torques which is usually provided to give stability at these conditions where the engine/rotor characteristics are different from those at the higher torque settings.

As an illustration of the effects of static droop, a simple mathematical model has been derived for the rotor speed variation due to a torque change in the rotor. (The slope of the basic engine law is denoted by $k$.)

If a change in rotor speed of $\delta\Omega$ occurs then a torque change of $-k \cdot \delta\Omega$ must occur according to the response characteristics of the engine to torque demands. If the response of the engine is $r(t)$, then at time $t$, the engine torque change triggered by the rotor speed variation $\delta\Omega$, at time $u$, is given by

$$\Delta Q = -k \cdot \delta\Omega \cdot r(t-u)$$
$$= -k \cdot \frac{d\Omega}{du} \cdot r(t-u) \cdot \delta u \tag{13.1}$$

hence the torque supplied by the engine at time $t$ is given by a superposition integral

$$\Delta Q = -k \int_0^t \frac{d\Omega}{du}(u) \cdot r(t-u) du \tag{13.2}$$

i.e. each infinitesimal rotor speed variation, at time $u$, will cause its own input to the engine torque which at time $t$ will have responded by $r(t-u)$.

Consider the initial steady-state; the engine input torque, $Q_N$, is balanced by a rotor torque demand $B$, i.e.

$$Q_N = B \tag{13.3}$$

if $B$ is then increased by an amount $\Delta B$, then the torque supplied to the rotor is

$$Q = Q_N - k\int_0^t \frac{d\Omega}{du}(u)\,r(t-u)\,du - (B+\Delta B) \qquad (13.4)$$

from which the equation of motion for the rotor speed is given by

$$I\frac{d\Omega}{dt} = -\Delta B - k\int_0^t \frac{d\Omega}{du}r(t-u)\,du \qquad (13.5)$$

if $\Omega_N$ is the initial steady-state rotor speed and $w(t)$ is the perturbation from it, then

$$\Omega = \Omega_N + w \qquad (13.6)$$

then the equation of motion for the rotor speed perturbation, becomes finally

$$I\frac{dw}{dt} = -\Delta B - k\int_0^t \frac{dw}{du}r(t-u)\,du \qquad (13.7)$$

The solution is easily accomplished using Laplace transforms.

Two cases are considered: (a) an engine with an immediate response and (b) an engine with a response defined by an exponential delay. The behaviour for both these situations is shown in Fig. 13.2 and is considered below.

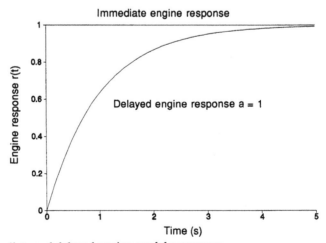

**Fig. 13.2.** Immediate and delayed engine model responses.

## Immediate engine response

In this case

$$r(t) = 1 \tag{13.8}$$

As already discussed the situation is defined as a rotor/engine governor system in equilibrium when a torque demand of $\Delta B$ is required of the rotor (say, a pilot's input of increased collective pitch). The variation of rotor speed relative to the original is observed to be

$$w(t) = \frac{-\Delta B}{k}\left[1 - e^{-(kt/I)}\right] \tag{13.9}$$

where $w(t)$ is the rotor speed perturbation, $\Delta B$ is the torque demand increase which displaces the system from its equilibrium state, $k$ is the slope of the droop law (i.e. torque/rotor speed) and $I$ is the rotational inertia of the rotor and transmission. The shape of the rotor speed perturbation as defined by Eqn (13.9) is shown in Fig. 13.3. The data for this are $I = 500$ kg m², $\Delta B = 10\,000$ N m and $k = 1000$ N m s/rad. This shows a smooth, overdamped transition from the original rotor speed to the final disturbed equilibrium value.

This shape of curve is not realistic since an immediate engine response is an idealised situation. However, the analysis has been included since it illustrates one important fact. That is the static droop (or final displacement of the rotor speed) is equal to $-\Delta B/k$ which is the quotient of the torque demand change and the slope of the droop characteristic, as illustrated in Fig. 13.1. Therefore, for a given change in rotor torque, a larger slope of the droop law is accompanied by a small change

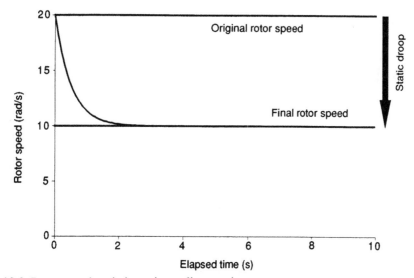

**Fig. 13.3.** Rotor speed variation—immediate engine response.

in rotor speed. This, at first sight, seems the behaviour required since variation in rotor speed must be minimised. However, the small change in rotor speed is bought at a price, and the second case of a delayed engine response will highlight what price that is.

### Delayed engine response

In this case the engine response is modified by the term

$$r(t) = 1 - e^{-at} \tag{13.10}$$

The response of this modified system can be obtained from the immediate engine response case using a superposition integral which assembles the engine torque changes over infinitesimal periods of time and sums the effect with the correct time delays. The equation of motion can be differentiated to give a linear second-order differential equation under an external forcing term. With such a system, the response can be under, critically, or overdamped and in order to give the type of rotor speed variation which typically occurs in practice, the underdamped solution is selected. Such a solution is of the form

$$w(t) = \frac{-\Delta B}{k}\left[1 + e^{(-at/2)}\left(\frac{2k - Ia}{I\varepsilon}\sin\left(\frac{\varepsilon t}{2}\right) - \cos\left(\frac{\varepsilon t}{2}\right)\right)\right]$$

$$\varepsilon^2 = \frac{4ka}{I} - a^2 \tag{13.11}$$

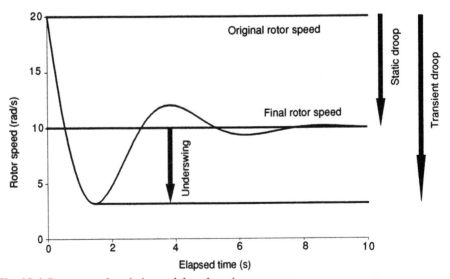

**Fig. 13.4.** Rotor speed variation—delayed engine response.

The value of $\varepsilon^2$ determines whether the response is under, critically, or overdamped. This variation of rotor speed is shown in Fig. 13.4, for the case $a = 1$.

The final rotor speed is displaced from the original value by the static droop value which again is equal to $-\Delta B / k$. The rotor speed, however, does drop below the static droop displacement and this maximum displacement is called the *transient droop*. The difference between the transient and static droops, or alternatively the maximum difference of the rotor speed below the final steady-state value, is called the *underswing*. In order to observe what effects the various governor parameters have on the rotor speed, the calculations were repeated for the following range of values: $\Delta B = 5000$, 7500, 10 000 N m, $a = 1$, 2, 3, $k = 1000$, 2000, 3000 N m s/rad. The italic values are those used in the basic case.

The effect on the rotor speed time-history of the various parameters is shown in Figs 13.5–13.7.

The conclusions drawn from Figs 13.5–13.7 are

(a) Increasing torque demand, $\Delta B$, does not change the character of the curves, only the final rotor speed drop which naturally increases with torque demand.

(b) Increasing the value of parameter $a$ means faster responding engine(s) and, as can be seen, the rotor speed will take a shorter time to settle for a faster responding engine.

(c) An increase in $k$ will decrease the final rotor speed perturbation as previously described. However, the behaviour is more variable and will take more overshoots to settle. This slow decay of the transient rotor speed behaviour will put greater stress on the transmission and rotor system (lag elements or dampers), and cause greater problems of fatigue.

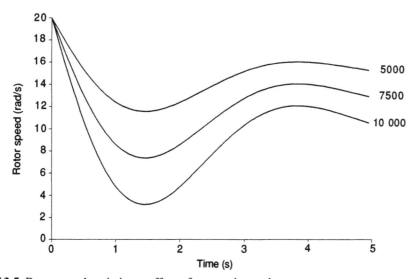

**Fig. 13.5.** Rotor speed variation—effect of torque demand.

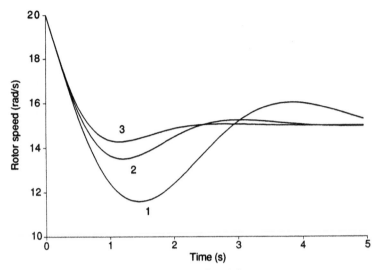

**Fig. 13.6.** Rotor speed variation—effect of engine delay.

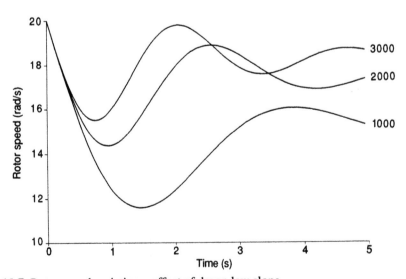

**Fig. 13.7.** Rotor speed variation—effect of droop law slope.

Therefore a large slope in the droop law will reduce the static droop and hence keep a tighter control on the rotor speed. However, the stability of the system is reduced and larger torsional forcings will occur in the system causing possible fatigue and vibration problems. This is the price which has to be paid for a close control of rotor speed. The benefits of high droop slope on rotor speed can be obtained without degrading the stability of the system, by altering the datum of the droop characteristic in a technique known as *static droop cancelling*, which is described below.

## Static droop cancelling

This technique is based on the requirement of a high droop slope (giving a small value of static droop and hence keeping variations in rotor speed to a minimum), without the disadvantages of potentially damaging torque variations in the transmission. With static droop cancelling, the droop law as shown in Fig. 13.1 is moved along the rotor speed abscissa as the steady-state torque demand changes. This movement of the droop law is controlled by a position signal from the collective pitch lever. The effect of this is shown in Fig. 13.8. (Adjustment is also possible for the part of the droop curves below the knee.)

From Fig. 13.8, the effect of static droop cancelling is to raise the slope of the droop law, as more torque is demanded by the pilot, as a result of raising the collective pitch lever. In this way the rotor speed variation is restricted to a small value. Any rotor speed perturbations not demanded by the pilot (encountering a wind gust for instance) will then adjust the engine throttle according to the basic droop law passing through the appropriate steady-state condition. As such a torque change is *not* accompanied by a movement of the collective pitch lever, it will be at the lower slope of the basic droop law and will therefore give a more stabilised control.

All parts of the droop law curves may not necessarily be affected by static droop cancelling. The pick-up point for this should, however, be below the minimum

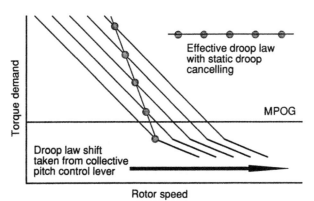

**Fig. 13.8.** Principle of static droop cancelling.

pitch on the ground (MPOG, an effective engine idling setting) and so will not affect normal powered flight.

## Engine matching

In a multi-engined helicopter, the matching of engine performances can be achieved by means of droop laws and static droop cancelling, and without the need for any inter-engine connections. The droop curve shown in Fig. 13.1 can be adjusted by means of static droop cancelling, above the static droop pick-up point as regards position and slope. In order to achieve a perfect match of engine torque versus engine rotational speed, the knee of the droop law curve must be below the MPOG level in order to achieve this over the whole power range. The reason for this is shown in Figs 13.9 and 13.10.

Figure 13.9 shows that with both knees below MPOG an engine match is possible over the complete powered flight range. Figure 13.10 shows that with only one knee below MPOG, alignment of the droop curves over the whole power range is impossible, hence perfect engine matching is not achievable.

The situation whereby *both* knees are above MPOG, engine matching is possible, however this change of slope is intended for use at low power and during autorotation and *not* for powered flight.

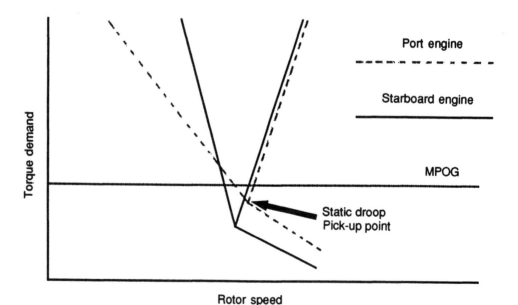

**Fig. 13.9.** Correctly matched engines.

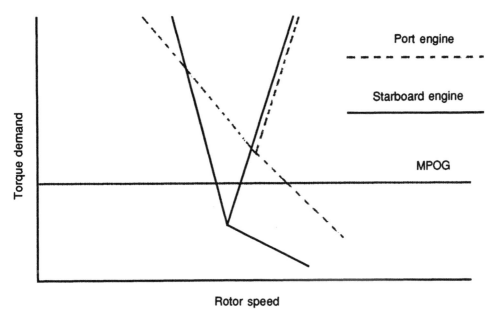

**Fig. 13.10.** Incorrectly matched engines.

# Governor failure

The engine system must naturally be subject to the chance of failure and the means of corrective action must be devised. A governor failure is normally assumed to produce one of the following three problems

(a) *An engine freeze*. If the aircraft is fitted with a manual reversion facility, then no problems should occur. Without this facility, the decision on further actions is dependent on the number of engines, the frozen power level and what landing facilities are immediately available?

(b) *An engine runaway up*. Upward engine runaways on a multi-engined helicopter are normally contained by an automatic reduction in power of the good engine(s). With a single-engined helicopter or a multi-engined helicopter at low power, the pilot is able to control the problem by increasing the collective pitch. If the pilot takes no action to control the situation, the appropriate free turbine speed trip will operate and shut the engine down. In this case the pilot must then handle an engine failure situation.

(c) *An engine runaway down*. Downward engine runaways can at most cause a situation very much like an engine failure. However, a manual reversion facility will allow a recovery to a safe flight condition if the loss in power does not constitute a hazard.

As well as an overall system failure, as described above, the failure may only affect a sub-assembly. The effect of such a failure is dependent on the failed system. Some system failures may only affect the transient characteristics, whilst others will seriously affect both the static and transient behaviour.

# 14

# Methods of calculating helicopter power, fuel consumption and mission performance

## Calculation of helicopter power and fuel consumption

The previous chapters have discussed methods to calculate the power required to fly a helicopter of a given weight at a given speed. The results of these calculations, together with engine data can enable the fuel consumption to be determined and hence a mission can be "flown". This enables a project study on a proposed helicopter design to be completed. The method to be described in this chapter collates the various momentum or actuator disc theories into a suggested calculation scheme. As explained in the earlier chapters, momentum theories are the simplest available and therefore require a degree of factoring to make the results more realistic. However, because of their simplicity they should not be applied for detailed rotor studies, particularly with respect to the flight envelope. With this proviso, the simplicity does in fact allow a parametric study to be achieved quickly and cheaply. With modern personal computers the methods described in this chapter can be readily handled and a broad picture of the helicopter design and its ability to perform a proposed mission can be quickly achieved.

This chapter describes a method of calculating the power required and fuel consumption of a helicopter. It is based on a momentum method used for project calculations and is only applicable for a general appraisal of the helicopter's performance and is designed for a single main and tail rotor configuration. It is unsuitable for investigating performance towards the boundaries of the aircraft's flight envelope.

The method is described in sections each dealing with a specific aspect of the helicopter. They are summarised as

(a)  The attitude of the main rotor disc and the thrust is calculated on the basis of it balancing the helicopter's drag and weight.
(b)  The main rotor, induced, profile, and parasite powers are then determined, using appropriate factors. These are summed to give the total power required to drive the main rotor.
(c)  The main rotor power is then converted to a torque which determines the tail rotor thrust required for yaw trim.
(d)  The tail rotor induced and profile powers can then be calculated.

(e) The total helicopter power required can then be determined by summing the main and tail rotor powers together with that required to drive auxiliary services.
(f) Transmission losses are then accounted for giving the power required *of the engines.*

## Notation

The terminology used in the analysis in this chapter is given in Tables 14.1 and 14.2.

**Table 14.1.** Rotor notation

|  | *Main* | *Tail* |
|---|---|---|
| Thrust | $T_M$ | $T_T$ |
| Blockage factor | $BLOCK_M$ | $BLOCK_T$ |
| Tip speed | $V_{TM}$ | $V_{TT}$ |
| No. of blades | $N$ | $n$ |
| Blade chord | $C$ | $c$ |
| Rotor radius | $R$ | $r$ |
| Disc tilt | $\gamma_s$ | — |
| Advance ratio | $\mu_M$ | $\mu_T$ |
| Advance ratio, component parallel to disc | $\mu_{xM}$ | $\mu_{xT}$ |
| Advance ratio, component perpendicular to disc | $\mu_{zM}$ | — |
| Downwash | $\lambda_{iM}$ | $\lambda_{iT}$ |
| Profile drag coefficient | $C_{D0M}$ | $C_{D0T}$ |
| Induced power factor | $k_{iM}$ | $k_{iT}$ |
| Induced power | $P_{iM}$ | $P_{iT}$ |
| Profile power | $P_{PM}$ | $P_{PT}$ |
| Parasite power | $P_{PARA\,M}$ | — |
| Total power | $P_{TOT\,M}$ | $P_{TOT\,T}$ |

**Table 14.2.** Overall helicopter notation

| | |
|---|---|
| Drag at 100 velocity units | $D_{100}$ |
| Auxiliary power | $P_{AUX}$ |
| Transmission loss factor | TRLF |
| Helicopter weight | $W$ |
| Helicopter drag | $D$ |

# Overall aircraft

## Calculate aircraft drag

Firstly the forward tilt of the main rotor needs to be established. This requires the estimation of the drag of the complete aircraft. Drag can be specified in many ways, however, this method uses as a basis the drag force at a reference speed of 100 units ($D_{100}$), at ISA sea level air density. The drag of the aircraft is then calculated using a variation of drag with the square of the forward speed and linearly with respect to the air density

$$D = D_{100}\left(\frac{V}{100}\right)^2 \sigma \qquad (14.1)$$

# Main rotor

## Calculate main rotor thrust and disc attitude

The force diagram for the main rotor is shown in Fig. 14.1, force balance requires resolving vertically

$$T\cos(\gamma_s) = W \qquad (14.2)$$

resolving horizontally

$$T\sin(\gamma_s) = D \qquad (14.3)$$

From which we obtain for the disc tilt

$$\gamma_s = \tan^{-1}\left(\frac{D}{W}\right) \qquad (14.4)$$

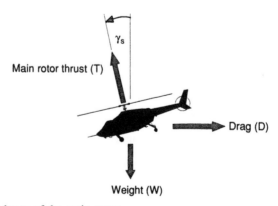

**Fig. 14.1.** Force balance of the main rotor.

The main rotor thrust, *with blockage*, becomes

$$T = \sqrt{W^2 + D^2} \times \text{BLOCK}_M \tag{14.5}$$

The blockage term is applied to the main rotor thrust as a multiplying factor and represents the download on the fuselage due to the rotor downwash.

The induced velocity of the main rotor is now required. As actuator disc theory is to be used, the main rotor advance ratio components parallel to and normal to the rotor disc plane are required together with the thrust coefficient.

The advance ratio is

$$\mu_M = \frac{V}{V_{TM}} \tag{14.6}$$

Resolving parallel to the rotor disc

$$\mu_{xM} = \mu_M \cos(\gamma_s) \tag{14.7}$$

Resolving perpendicular to the rotor disc

$$\mu_{zM} = \mu_M \sin(\gamma_s) \tag{14.8}$$

## Calculate main rotor downwash

The main rotor downwash can now be calculated using the iterative technique

$$\lambda_{iM\,\text{NEW}} = \lambda_{iM\,\text{OLD}} - \left[ \frac{\lambda_{iM\,\text{OLD}} - \dfrac{C_T}{4\sqrt{\mu_{xM}^2 + (\mu_{zM} + \lambda_{iM\,\text{OLD}})^2}}}{1 + \dfrac{(\mu_{zM} + \lambda_{iM\,\text{OLD}})C_T}{4\sqrt{\mu_{xM}^2 + (\mu_{zM} + \lambda_{iM\,\text{OLD}})^2}^3}} \right] \tag{14.9}$$

A sensible starting value is that for hover, namely

$$\lambda_{i\,\text{OLD}} = \frac{1}{2}\sqrt{C_T} \tag{14.10}$$

This will give the downwash $\lambda_{iM}$.

### Assemble main rotor powers

The main rotor power components are now calculated.
Induced rotor power (note the inclusion of the induced power factor $k_{iM}$).

$$P_{iM} = k_{iM} \cdot T_M \cdot V_{TM} \cdot \lambda_{iM} \qquad (14.11)$$

Profile rotor power

$$P_{pM} = \tfrac{1}{8} \rho (V_{TM})^3 \cdot NCR \cdot C_{D0M} \cdot (1 + 3.0 \mu_{xM}^2) \qquad (14.12)$$

Parasite rotor power

$$P_{PARA\,M} = D \cdot V \qquad (14.13)$$

Total rotor power

$$P_{TOT\,M} = P_{iM} + P_{pM} + P_{PARA\,M} \qquad (14.14)$$

It should be noted that the $\mu_x^2$ term in Eqns (14.12) and (14.19) is multiplied by 3.0. This is the value which was obtained in (5.40). The extra effects raising this to 4.7 can be simply dealt with by amending the calculation method.

## Tail rotor

The main rotor power is now converted to a torque from which the required tail rotor thrust for torque balance is calculated.

### Calculate tail rotor thrust

We calculate this as

$$T_T = \left( \frac{P_{TOT\,M}}{\Omega_M \cdot l_{BOOM}} \right) BLOCK_T \qquad (14.15)$$

The induced velocity of the tail rotor is also required, the following are then calculated

$$\mu_{xT} = \frac{V}{V_{TT}} \qquad (14.16)$$

$$\mu_{zT} = 0 \qquad (14.17)$$

Since the tail rotor is only required to develop a thrust with no disc tilt, the disc plane is assumed parallel to the flight path.

### Calculate tail rotor downwash

Using the same iterative method, the tail rotor downwash, $\lambda_{iT}$, can be calculated.

### Assemble tail rotor powers

The tail rotor power components can now be calculated.

Induced

$$P_{iT} = k_{iT} \cdot T_T \cdot V_{TT} \cdot \lambda_{iT} \qquad (14.18)$$

Profile

$$P_{pT} = \tfrac{1}{8} \rho (V_{TT})^3 \cdot ncr \cdot C_{D0T} \cdot (1 + 3.0\mu_{xT}^3) \qquad (14.19)$$

Total

$$P_{\text{TOT}\,T} = P_{iT} + P_{pT} \qquad (14.20)$$

Note that the main rotor is responsible for overcoming the parasite drag of the aircraft.

## Complete aircraft

The total power required can now be calculated by summing the total powers for each rotor together with any auxiliary services $(P_{\text{AUX}})$ such as oil pumps and electrical generators. Since some losses will occur in the transmission, a factor is applied to account for this giving the power required from the engines. The formula is

$$P_{\text{TOTAL}} = [P_{\text{TOT}\,M} + P_{\text{TOT}\,T} + P_{\text{AUX}}] \cdot \text{TRLF} \qquad (14.21)$$

## Example of parameter values

Table 14.3 gives example values of the various factors applied.

**Table 14.3**

| Parameter | Symbol | Main rotor | Tail rotor |
|-----------|--------|------------|------------|
| Rotor blockage | BLOCK | 1.05 | 1.10 |
| Induced power factor | $k_i$ | 1.10 | 1.20 |
| Profile drag coefficient | $C_{p0}$ | 0.011 | 0.012 |

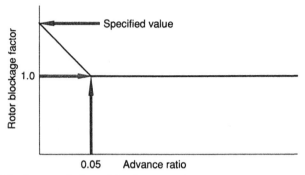

**Fig. 14.2.** Rotor blockage variation with advance ratio.

The blockage effect on each rotor will vary with forward speed. The main rotor blockage is a result of the main rotor downwash impinging on the fuselage causing a download. Therefore, in order to achieve the required component of force from the rotor, the thrust must exceed this by an amount equal to the download on the fuselage. The blockage factor multiplies the *desired net thrust* to account for the loss, however, as the forward speed of the helicopter increases, the rotor downwash is carried increasingly downstream of the fuselage eventually causing no real fuselage download from the main rotor wake. In this case the blockage factor becomes unity. The variation specified for this example is based on advance ratio and the blockage value refers to the hover condition and linearly decreases to unity at an advance ratio of 0.05. Thereafter it remains at unity as shown in Fig. 14.2.

## Calculation of engine fuel consumption

Having determined the power required from the engine(s) to fly a given helicopter flight condition, the fuel consumption can now be calculated.

Engine fuel consumption data are often given in terms of specific fuel consumption (SFC) (kg/h/kW) for a corresponding power setting ($P$). A small conversion of the data allows a very simple method of calculating the fuel consumption of a gas turbine engine developing a specified amount of power. The fuel flow of an engine ($W_f$) (kg/h) can be obtained as the product of the SFC and the respective power. Plotting fuel flow against power produces a variation very close to linear which can be fitted by means of linear regression (least squares). The fuel consumption calculation becomes elementary because of this simple linear variation.

To enable full use to be made of the method, the operating altitude and temperature will need to be considered in the calculation. Each atmospheric condition will produce an individual straight line fit, but if the fuel flow and engine power are normalised by $\delta\sqrt{\theta}$, where $\delta$ is the pressure ratio and $\theta$ is the absolute temperature ratio (both relative to ISA sea level atmosphere conditions) the various lines defining the engine performance at a given atmospheric condition close together and the straight line law for the fuel flow becomes practically one line. That is the engine fuel consumption law for any atmospheric condition becomes

$$\frac{W_f}{\delta \sqrt{\theta}} = A_E + B_E \left[ \frac{P}{\delta \sqrt{\theta}} \right] \qquad (14.22)$$

This gives complete flexibility as regards changing atmospheric conditions. It should be noted that this type of engine calculation assumes that the power unit is not operating close to a limit.

The straight line fit possesses a positive intercept on the fuel flow axis $(A_E)$ which results in an important fact concerning the optimising of fuel consumption for a multi-engined helicopter.

Consider a helicopter which is fitted with $N$ engines combining to give a total power production of $P$. Each engine must then produce a power of $P/N$ and the corresponding fuel consumption for all $N$ engines is therefore

$$\frac{W_f}{\delta \sqrt{\theta}} = N \left( A_E + B_E \left[ \frac{P}{N} \times \frac{1}{\delta \sqrt{\theta}} \right] \right) \qquad (14.23)$$

i.e.

$$W_f = N \cdot A_E \cdot \delta \sqrt{\theta} + B_E \cdot P \qquad (14.24)$$

It can be seen that for a given power production, because of the first term of the right-hand side of (14.24), a smaller number of engines will give a lower fuel consumption. Consequently a helicopter designer, when optimising for fuel consumption, will use a minimum number of engines capable of providing sufficient power. Other requirements such as helicopter performance with an engine failure tend to work in the opposite direction and on this type of basis, the optimum design will have a maximum number of engines. Hence, choice of engine for a multi-engine helicopter design is not as clear cut as might at first be thought.

## Engine limits

The previous discussions have considered the engine performance purely from a fuel consumption point of view. In reality, a gas turbine engine will have limitations which are usually based on the permissible operating temperature of the turbine section. However, it is possible to operate the engine at higher power settings for a limited period of time without causing permanent damage. In emergency situations, some excessive power settings can be demanded for very limited time periods at the expense of accelerated engine wear or damage. Typical examples of power settings are described below.

- *Maximum continuous power rating.* This is the maximum power that an engine can develop without a time constraint and consequently operate continuously.

- *Take-off or one-hour power rating.* This is used for the higher power conditions such as take-off and hover at high altitude and/or ambient temperature. It can be used for periods of approximately one hour (sometimes half an hour) before the power demand must be reduced.
- *Maximum contingency or $2\frac{1}{2}$-minute power rating.* This is used with engine loss and other contingency situations when the power can be used for short periods of 2–3 minutes. An engine inspection is usually considered after this type of use.
- *Emergency or half-minute power rating.* This is the highest and consequently the most damaging condition. It is used for situations where loss of the aircraft is a real possibility.

An example is when a twin-engined naval helicopter suffers an engine loss in a high power condition (high all-up weight and in the hover) and is forced to ditch in the sea. After jettisoning as much weight as possible, a take-off on a single engine from the water will be necessary to save the aircraft. Emergency power will be required for such a dire emergency and the engine will probably require extensive maintenance and refurbishment, if not scrapping, due to the use of such an excessive, but necessary, power demand.

## Calculation of the performance of a helicopter

As an illustration the following calculations, based on a small, twin-engined utility helicopter, are presented. The input data required are shown in Tables 14.4 and 14.5. The fuel flow law for each engine, presented in (14.25), which was obtained from public domain information, using linear regression is

$$\frac{W_f}{\delta\sqrt{\theta}} = 46.5 + 0.24\frac{P}{\delta\sqrt{\theta}} \tag{14.25}$$

With the data in Tables 14.3 and 14.4, the variation of the main rotor power components with forward speed is shown in Fig. 14.3, namely induced, profile, parasite and total.

It is instructive to see these results when shown cumulatively as in Fig. 14.4.

**Table 14.4**

| Rotor data | Main rotor | Tail rotor |
|---|---|---|
| No. of blades | 4 | 4 |
| Chord (m) | 0.394 | 0.180 |
| Radius (m) | 6.4 | 1.105 |
| Tip speed (m/s) | 218.69 | 218.69 |
| Blockage | 1.05 | 1.10 |
| Induced power factor $(k_i)$ | 1.10 | 1.20 |
| Profile drag coefficient $(C_{D0})$ | 0.011 | 0.012 |

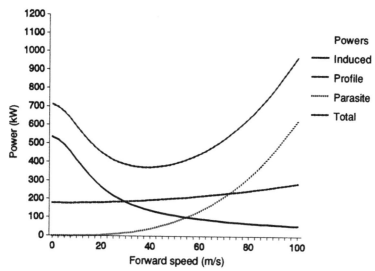

**Fig. 14.3.** Variation of main rotor power components in forward flight.

**Table 14.5.** Fuselage data

| | |
|---|---:|
| Tail boom length (m) | 7.66 |
| $D_{100}$ (N) | 6226.9 |
| Auxiliary power $P_{AUX}$ (kW) | 26.1 |
| Transmission loss factor (TRLF) | 1.04 |

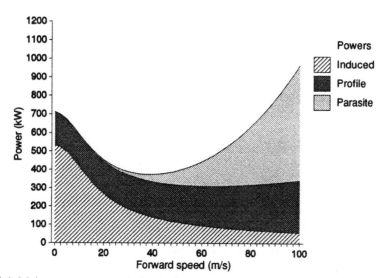

**Fig. 14.4.** Main rotor power variation in forward flight.

The following are then added to the main rotor power: tail rotor power, auxiliary power and the influence of transmission losses to give the total overall power required of the powerplants. The total power variation of the complete helicopter with forward speed is shown in Fig. 14.5.

From this power distribution the fuel flow can now be calculated and converted to endurance and range estimates. Endurance is the time required to consume a specified amount of fuel, whilst range is the corresponding distance covered. Endurance is concerned principally with minimising the rate of fuel usage with time, while range is a compromise between time and forward speed. (A fuel usage of 100 kg is assumed for these calculations.) The endurance is shown in Fig. 14.6, and the range (km) is shown in Fig. 14.7. Two plots are presented in each figure corresponding to the full fuel flow law defined in Eqn (14.25), and a modified law where the intercept ($A_E$, or fuel flow at zero power) is set to zero. (This corresponds to a constant SFC.) The upper curve corresponds to the full law ($A_E \neq 0$) and the lower curve to the modified law ($A_E = 0$).

The positions of maximum endurance and range are indicated in the figures. As can be seen, changing the fuel flow law does not alter the optimum endurance speed of 38 m/s (A), which corresponds to minimum power. However, the change does have an influence on the optimum range speed increasing it from 65 m/s (B) to 80 m/s (C). Figure 14.8 shows the three extrema indicated by A, B and C on the total power vs forward speed curve. In fact, the three points A, B and C, corresponding to the various extrema, can be determined geometrically as the following analysis shows. The following definitions are used: $W_{FUEL}$ is the fuel weight (fixed); $T$, time; $S$, SFC; $P$, power; $V$, forward speed; $E$, endurance; and $R$, range.

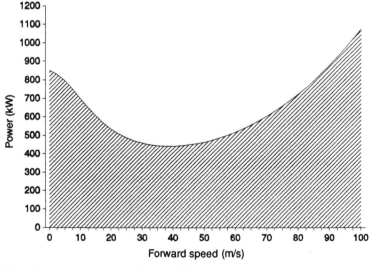

**Fig. 14.5.** Total power variation in forward flight.

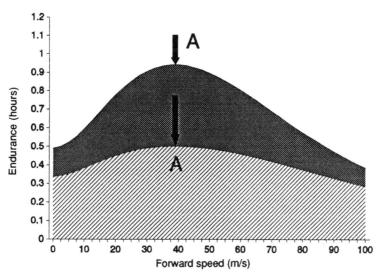

**Fig. 14.6.** Endurance variation in forward flight (100 kg of fuel).

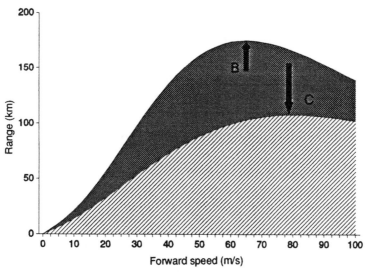

**Fig. 14.7.** Range variation in forward flight (100 kg of fuel).

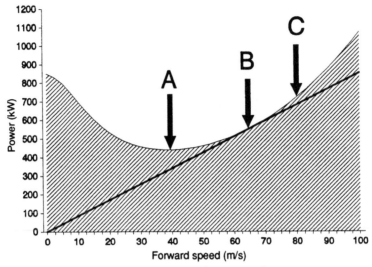

**Fig. 14.8.** Location of endurance and range maxima on total power curve.

## Point A (maximum endurance)

The definition for SFC is

$$S = \frac{W_{FUEL}}{P \times T} \qquad (14.26)$$

i.e.

$$T = \frac{W_{FUEL}}{S} \times \frac{1}{P} \qquad (14.27)$$

As $W_{FUEL}$ and $S$ are fixed, $T$ is maximum when $P$ is minimum, i.e. point A is the minimum point on the power vs velocity curve.

## Point B (maximum range—constant SFC, $A_E = 0$)

The range is given by

$$R = T \times V$$

$$= \frac{V}{P} \times \frac{W_{FUEL}}{S} \qquad (14.28)$$

i.e. $R$ is a maximum when $P/V$ is a minimum, so point B is positioned where the tangent, drawn from the origin, touches the power curve.

## Point C (maximum range—full fuel flow law)

For this case we have

$$W_{\text{FUEL}} = T(A_E + B_E P) \tag{14.29}$$

Therefore

$$R = V \times T$$

$$= \frac{V}{A_E + B_E P} \times W_{\text{FUEL}} \tag{14.30}$$

$$= \frac{W_{\text{FUEL}}}{B_E} \times \frac{V}{(A_E / B_E) + P}$$

This is very similar to point B except that the additional term in the denominator $(A_E / B_E)$ requires that the tangent be drawn from the point $(0, -A_E / B_E)$. The construction for the three points is shown in Fig. 14.9.

# Mission analysis

The calculation of engine power and fuel consumption can be extended to "fly" a mission in a computer which is of direct use to a project assessment. In flying a mission, the helicopter weight will be changing due to fuel usage and consequently this must be accounted for in the calculations. The method described here uses a technique in which this is simple to implement. The mission is divided into legs,

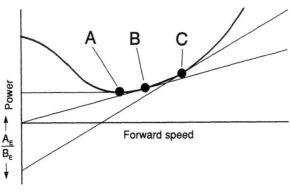

**Fig. 14.9.** Graphical construction of performance extrema.

each being calculated separately, in sequence, with the value of the aircraft weight at the end of a particular leg becoming the start value for the succeeding leg. (Each leg will describe a fixed flight condition, particularly altitude and/or forward speed.) This also allows discrete weight changes of payload to be incorporated, such as changes in passenger/cargo payload or the deployment of ordnance.

For each leg of the mission the power and fuel consumption at the start weight is calculated. Knowing the duration of the leg a first estimate of the weight change over the leg is obtained. Taking the mean value of aircraft weight the process is repeated and the fuel usage of the leg at the two weights compared. If they lie within a specified tolerance, the final estimate is taken. If not, a revised mean aircraft weight is adopted (using the latest fuel usage figure) and the process repeated iteratively until convergence to within the required tolerance is achieved.

# Calculation method

A summary of the calculation method is presented and can be used as a basis for a flow diagram. Each part of the calculation is presented as a complete entity with the input and output data being specified. This imparts a segmented structure to the method which is recommended for implementation in a computer program.

## *Atmospheric parameters*

This allows the air characteristics required by the calculations to be determined from the altitude and ambient temperature. (The helicopter is assumed to remain within the troposphere.)

1.  Input: altitude, sea-level air temperature, sea-level air density.
2.  Output: density ratio ($\sigma$), absolute temperature ratio ($\theta$), pressure ratio ($\delta$).
3.  Calculation: the absolute temperature ratio is given by

$$\theta = \frac{T_{\text{sea level}} - \text{altitude} \times \text{lapse rate}}{T_{\text{sea level}}} \qquad (14.31)$$

The pressure ratio by

$$\delta = \theta^{5.256} \qquad (14.32)$$

The relative density by

$$\sigma = \theta^{4.256} \qquad (14.33)$$

The lapse rate of 6.5°C per kilometre is used in the calculations.

## Downwash calculation

This is the application of momentum theory to the downwash calculation.

1. Input: thrust coefficient $(C_T)$, advance ratio components, parallel to the rotor disc $(\mu_x)$ and perpendicular to the rotor disc $(\mu_z)$.
2. Output: downwash $(\lambda_i)$.
3. Calculation:

   start value (hover)

$$\lambda_{i\,\text{OLD}} = \frac{1}{2}\sqrt{C_T} \qquad (14.34)$$

†

4.
$$\Delta\lambda_i = \left[\frac{\lambda_{i\,\text{OLD}} - \dfrac{C_T}{4\cdot\sqrt{\mu_x^2 + (\mu_z + \lambda_{i\,\text{OLD}})^2}}}{1 + \dfrac{(\mu_z + \lambda_{i\,\text{OLD}})C_T}{4\cdot[\mu_x^2 + (\mu_z + \lambda_{i\,\text{OLD}})^2]^{1.5}}}\right] \qquad (14.35)$$

$$\lambda_{i\,\text{NEW}} = \lambda_{i\,\text{OLD}} - \Delta\lambda_i \qquad (14.36)$$

5. Has the iteration reached convergence?

$$\left|\lambda_{i\,\text{NEW}} - \lambda_{i\,\text{OLD}}\right| < \text{tolerance?} \qquad (14.37)$$

6. If no then reset the downwash value and calculate the next estimate

$$\lambda_{i\,\text{OLD}} = \lambda_{i\,\text{NEW}} \qquad (14.38)$$

Go to †.

7. If yes then convergence has been achieved

$$\text{downwash} = \lambda_{i\,\text{NEW}} \qquad (14.39)$$

8. Then exit.

## Helicopter power

This is the helicopter power calculation.

1. Input: aircraft all-up weight, forward speed, atmospheric data, aircraft data. Suffix $M$ refers to the main rotor and $T$ refers to the tail rotor.
2. Output: helicopter power.
3. Calculation:

fuselage drag and aircraft weight

$\Downarrow$

main rotor thrust (with blockage added), disc tilt and forward speed

$\Downarrow$

$C_{TM}, \mu_{xM}, \mu_{zM}$

$\Downarrow$

$\lambda_{iM}$

$$\begin{bmatrix} P_{iM} \\ +P_{pM} \\ +P_{PARAM} \\ \hline P_{TOTM} \end{bmatrix} \tag{14.40}$$

$$\text{net tail rotor thrust} = \begin{bmatrix} \dfrac{\text{main rotor torque}}{\text{tail boom length}} \end{bmatrix} \tag{14.41}$$

4. Tail rotor thrust (with blockage added and assuming no disc tilting) and forward speed

$\Downarrow$

$C_{TT}, \mu_{xT}, (\mu_{zT} = 0)$

$\Downarrow$

$\lambda_{iT}$

$$\begin{bmatrix} P_{iT} \\ +P_{pT} \\ \hline P_{\text{TOT }T} \end{bmatrix}$$

(14.42)

$$\begin{bmatrix} P_{\text{TOT }M} \\ +P_{\text{TOT }T} \\ +P_{\text{AUX}} \\ \hline P_{\text{TOTAL}} \end{bmatrix}$$

(14.43)

5.      power required = transmission loss factor (TRLF) $\cdot P_{\text{TOTAL}}$      (14.44)

6.  Then exit.

## Fuel flow

This calculates the fuel consumption rate from the power requirements.

1.  Input: engine power ($P$), number of engines ($N$), atmospheric pressure ratio ($\delta$), atmospheric temperature ratio ($\theta$), engine performance coefficients ($A_E$, $B_E$).
2.  Output: fuel flow for the stated power required ($W_f$).
3.  Calculation:

$$\text{fuel flow } (W_f) = \delta \sqrt{\theta} \cdot N \cdot A_E + P \cdot B_E$$

(14.45)

4.  Then exit.

## Mission leg

This calculates the fuel usage over a mission component or leg.

1.  Input: start and finish altitudes (power and fuel flow are averaged between these two conditions), start all-up weight ($\text{AUW}_{\text{START}}$), forward speed ($V$), time or distance of leg (e.g. 5 min hover or 20 km cruise at 100 m/s).
2.  Output: fuel used during mission leg.
3.  Calculation: set variables to the conditions at the start of the leg

$$\text{AUW}_{\text{START}}, \ V$$

$$\Downarrow$$

calculate power

$$\Downarrow$$

calculate fuel flow

$$\Downarrow$$

calculate fuel usage = fuel flow $\times$ time $\Rightarrow$ fuel used$_{\text{OLD}}$

$$\Downarrow$$

calculate average weight = AUW = $\text{AUW}_{\text{START}} - \frac{1}{2}$ fuel used$_{\text{OLD}}$

$$\Downarrow$$

$$\dagger$$

4.   Calculate the fuel usage with the revised helicopter weigh

$$\text{AUW}, \ V$$

$$\Downarrow$$

power

$$\Downarrow$$

fuel flow

$$\Downarrow$$

fuel flow $\times$ time $\Rightarrow$ fuel used$_{\text{NEW}}$

5.            Has the fuel usage estimate converged?

$$\left| \text{fuel used}_{\text{OLD}} - \text{fuel used}_{\text{NEW}} \right| < \text{tolerance?} \qquad (14.46)$$

6.   If no, then reset the helicopter weight and proceed to a new estimate

$$\text{fuel used}_{\text{OLD}} = \text{fuel used}_{\text{NEW}}$$

$$\Downarrow$$

$$\text{AUW} = \text{AUW}_{\text{START}} - \frac{1}{2} \ \text{fuel used}_{\text{NEW}}$$

$$\Downarrow$$

go to $\dagger$

7.  If yes, then fuel usage calculation has converged and therefore the weight change over the leg can be determined

$$AUW_{FINISH} = AUW_{START} - fuel\ used_{NEW}$$

8.  Then exit.

## Examples of mission calculations

To see the application of the calculation to a mission, two examples are presented here. They are for the WG13 aircraft flying: (i) a mission performed mainly at high speed, such as anti-tank and (ii) a mission containing much hover time, such as anti-submarine (ASW). They do not represent existing missions but are used merely to examine the performance of a given helicopter engaged on missions which contain either significant elements of high speed operation or hover time.

The mission calculation was performed using the Westland WG13/Lynx as a datum aircraft for five helicopter designs, each illustrating the type of parametric change which a designer could choose. The results show the effect of such changes on the fuel usage of the Lynx aircraft. The five designs are listed below:

1.  This is the basic aircraft.
2.  This has the aircraft parasitic drag doubled via the $D_{100}$ term.
3.  This has the main and tail rotor radii increased by 0.5 m, with the consequent increase in tail boom length of 1 m.
4.  The number of engines are reduced to one. No allowance has been made for the ability of the single engine to generate sufficient power. Only the fuel consumption has been studied.
5.  The number of engines are increased to three. The data used for the power calculation of the Westland Lynx are used as a basis for these calculations.

The five sets of parameter changes are given in Table 14.6.

**Table 14.6.** Designs used for mission calculations

| Design No. | 1 | 2 | 3 | 4 | 5 |
|---|---|---|---|---|---|
| $D_{100}$ | 6227 | 12454 | 6227 | 6227 | 6227 |
| Main rotor radius | 6.401 | 6.401 | 6.901 | 6.401 | 6.401 |
| Tail rotor radius | 1.105 | 1.105 | 1.605 | 1.105 | 1.105 |
| Tail boom length | 7.66 | 7.66 | 8.66 | 7.66 | 7.66 |
| No. of engines | 2 | 2 | 2 | 1 | 3 |

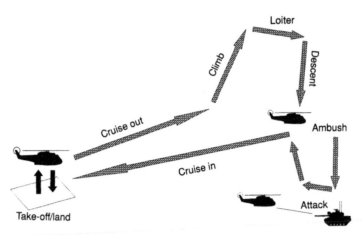

**Fig. 14.10.** Schematic of the anti-tank mission.

## Westland Lynx—anti-tank mission

The description of the mission is shown in Fig. 14.10. The mission is divided into nine distinct legs which are shown in Table 14.7.

**Table 14.7.** Anti-tank mission

| Leg | Phase | Altitude | Speed | Time | Distance | Remarks |
|-----|-------|----------|-------|------|----------|---------|
| 1 | Take-off | Sea level | Hover | 5 min | | |
| 2 | Cruise | Sea level | 70 m/s | | 100 km | |
| 3 | Climb | Sea level –2500 m | 50 m/s | 2 min | | |
| 4 | Loiter | 2500 m | 35 m/s | 15 min | | |
| 5 | Descent | 2500 m– sea level | 55 m/s | 10 min | | |
| 6 | Ambush | Sea level | Hover | 5 min | | |
| 7 | Attack | Sea level | 80 m/s | 5 min | | Drop 130 kg of ordnance |
| 8 | Return | Sea level | 70 m/s | | 100 km | |
| 9 | Land | Sea level | Hover | 5 min | | |

*Progression of calculation.* The method commences with the helicopter at the start of leg 1, at its take-off weight of 4500 kg. A tolerance of 5 kg of fuel usage for each leg is used. Table 14.8 shows the results.

**Table 14.8.** Leg 1—hover 5 min (0.083 h)

| Pass No. | Aircraft weight (kg) | Power (kW) | Fuel flow (kg/h) | Fuel used (kg) | Average aircraft weight (kg) |
|----------|----------------------|------------|------------------|----------------|------------------------------|
| 1 | 4500 | 949 | 322 | 27 | 4487 |
| 2 | 4487 | 946 | 321 | 27 | |

The "fuel used" calculation has therefore converged after the second pass at 27 kg. Aircraft weight at the end of the first leg and hence the beginning of the second leg is

$$4500 - 27 = 4473 \text{ kg}$$

The calculation of the second leg gives the results shown in Table 14.9.

**Table. 14.9.** Leg 2—cruise 100 km at 70 m/s for 24 min (0.397 h)

| Pass No. | Aircraft weight (kg) | Power (kW) | Fuel flow (kg/h) | Fuel used (kg) | Average aircraft weight (kg) |
|---|---|---|---|---|---|
| 1 | 4473 | 620 | 242 | 96 | 4425 |
| 2 | 4425 | 617 | 242 | 96 | |

The "fuel used" calculation has converged after the second pass at 96 kg. Aircraft weight at the end of the second leg and hence the beginning of the third leg is

$$4473 - 96 = 4377 \text{ kg}$$

This calculation now proceeds through the remaining seven legs of the mission.

*Fuel consumption.* The summary of the fuel consumption for all nine mission legs and for each of the five helicopter configurations is given in Table 14.10. The total fuel consumption is included, expressed as a mass and as a percentage relative to the basic helicopter (design 1).

**Table 14.10.** Anti-tank mission

| Mission leg | 1 | 2 | Design No. 3 | 4 | 5 |
|---|---|---|---|---|---|
| 1 | 27 | 27 | 26 | 23 | 31 |
| 2 | 96 | 118 | 97 | 77 | 114 |
| 3 | 14 | 15 | 14 | 13 | 16 |
| 4 | 45 | 46 | 43 | 37 | 53 |
| 5 | 25 | 29 | 25 | 18 | 31 |
| 6 | 26 | 26 | 25 | 22 | 29 |
| 7 | 22 | 29 | 23 | 19 | 26 |
| 8 | 94 | 116 | 96 | 76 | 113 |
| 9 | 24 | 24 | 24 | 21 | 28 |
| Total | 373 | 430 | 373 | 306 | 441 |
| Percentage of design 1 | 100 | 115 | 100 | 82 | 118 |

As can be seen in design 2, changes in $D_{100}$ increase the fuel consumption by 15% consistent with the amount of time spent at high speed where parasite power dominates. Changes to rotor size, in design 3, produce virtually no change in fuel consumption. This is explained by the fact that an increase in disc area reduces the hover power (induced component), but the attendant increase in blade area increases the profile power, particularly at high speed, and for this mission the two effects cancel out. The changes in the number of engines, designs 4 and 5, are apparently the most influential, showing the advantage, from a fuel consumption viewpoint, of operating on a single engine, and on the same basis, the disadvantage of carrying a third engine.

These results are plotted in Figs 14.11–14.16. Figure 14.11 shows the fuel usage for the nine legs and under each leg, the five helicopter configurations. Figures 14.12–14.16 show the helicopter weight variation with time for each configuration over the complete mission.

## Westland Lynx—ASW mission

As a comparison, a second mission is presented. The anti-tank mission previously discussed uses high speed to a great degree. As was seen in the results of the anti tank mission, the influence of rotor radius is not particularly influential in this mission which consists mainly of conditions of high speed where parasite power is dominant. The ASW mission, described below, however contains much hover time and so will illustrate the importance of rotor radius for this type of role. The description of the mission is shown in Fig. 14.17. As before the mission is divided into legs which are shown in Table 14.11. With the exception of increasing the

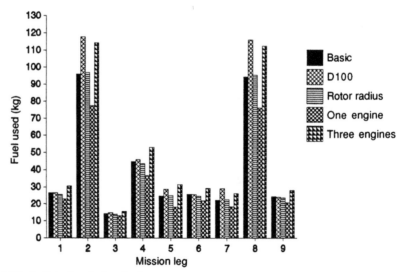

**Fig. 14.11.** Anti-tank mission fuel usage.

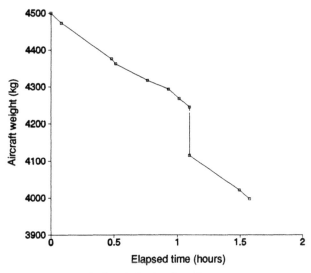

**Fig. 14.12.** Anti-tank mission, design 1: basic aircraft.

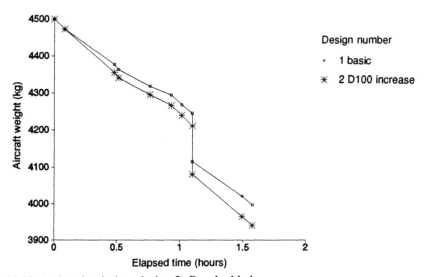

**Fig. 14.13.** Anti-tank mission, design 2: $D_{100}$ doubled.

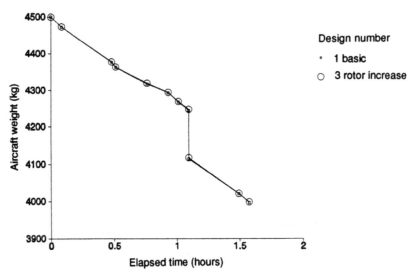

**Fig. 14.14.** Anti-tank mission, design 3: rotor radii +0.5 m, tail boom +1 m.

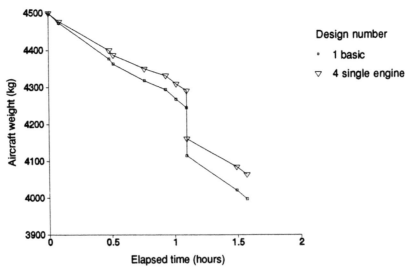

**Fig. 14.15.** Anti-tank mission, design 4: single engine.

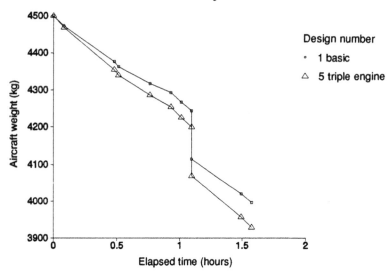

**Fig. 14.16.** Anti-tank mission, design 5: three engines.

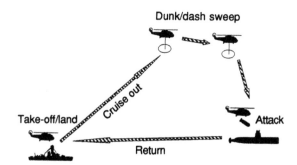

**Fig. 14.17.** Schematic of the ASW mission.

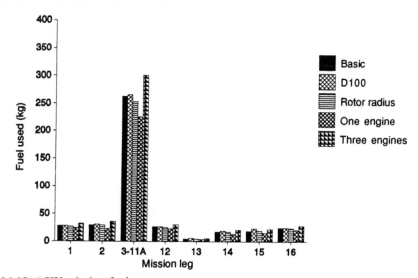

**Fig. 14.18.** ASW mission fuel usage.

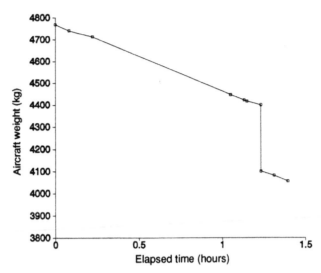

**Fig. 14.19.** ASW mission, design 1: basic aircraft.

start AUW from 4500 to 4770 kg, the aircraft data and the parameter changes are identical to the anti-tank mission.

*Fuel consumption.* The fuel consumption for all 16 mission legs and the five configurations is in Table 14.12 and in Figs 14.18–14.23.

**Table 14.11.** ASW mission

| Leg | Phase | Altitude | Speed | Time | Distance | Remarks |
|-----|-------|----------|-------|------|----------|---------|
| 1 | Take-off | Sea level | Hover | 5 min | | |
| 2 | Cruise | Sea level | 40 m/s | | 20 km | |
| 3A | Dunk | Sea level | Hover | 5 min | | The dunk, dash sequence |
| 3B | Dash | Sea level | 60 m/s | | 2 km | is performed nine times. 3A–11B |
| 12 | Dunk | Sea level | Hover | 5 min | | |
| 13 | Dash | Sea level | 80 m/s | | 5 km | |
| 14 | Attack | Sea level | 50 m/s | 5 min | | Drop 300 kg of torpedoes |
| 15 | Return | Sea level | 70 m/s | | 20 km | |
| 16 | Land | Sea level | Hover | 5 min | | |

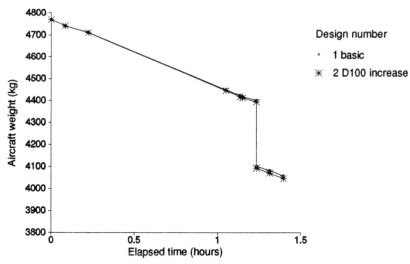

**Fig. 14.20.** ASW mission, design 2: $D_{100}$ doubled.

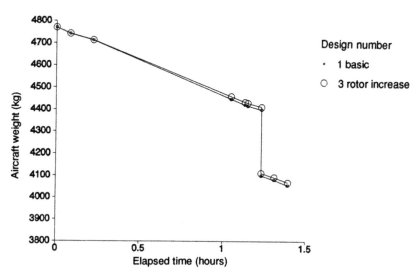

**Fig. 14.21.** ASW mission, design 3: rotor radii +0.5 m, tail boom +1 m.

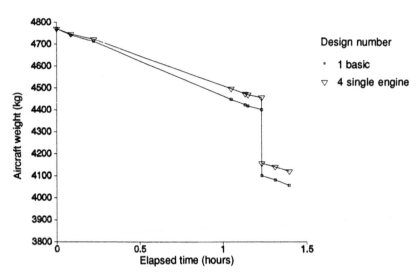

**Fig. 14.22.** ASW mission, design 4: single engined.

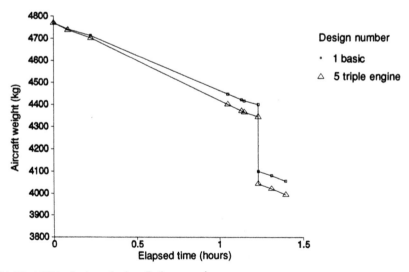

**Fig. 14.23.** ASW mission, design 5: three engines.

**Table 14.12.** ASW mission

| Mission leg | Design No. | | | | |
|---|---|---|---|---|---|
| | 1 | 2 | 3 | 4 | 5 |
| 1 | 28 | 28 | 27 | 24 | 32 |
| 2 | 29 | 31 | 29 | 23 | 36 |
| 3A–11B | 264 | 266 | 253 | 226 | 300 |
| 12 | 26 | 26 | 25 | 23 | 30 |
| 13 | 5 | 6 | 5 | 4 | 5 |
| 14 | 17 | 19 | 17 | 14 | 21 |
| 15 | 19 | 23 | 19 | 15 | 23 |
| 16 | 25 | 25 | 24 | 21 | 28 |
| Total | 413 | 424 | 399 | 350 | 475 |
| Percentage of design 1 | 100 | 103 | 97 | 85 | 115 |

The effect of the change in $D_{100}$ (design 2) is reduced from the anti-tank mission because the mission is primarily concerned with low-speed hovering flight. This reason also accounts for the 3% advantage gained from increasing the size of the rotors and tail boom (design 3). The arguments about the number of engines still apply.

In summary, the parasite drag has a great influence on the performance if the mission contains substantial periods of high speed. This is particularly true if external stores are being carried. Increasing rotor size has a beneficial effect if hover forms an important part of a mission. However, should the blade chord remain unaltered, then the increase in blade area will cause an increase in profile power and hence work in opposition to the benefits gained from the lower disc loading. The net effect of rotor size will therefore depend on the balance between low and high speed operation throughout the particular mission.

The effect of the number of engines on fuel consumption is of major importance.

# Epilogue—the windmill

This book has attempted to introduce the reader to the basic aerodynamic and dynamic features of the helicopter. At the beginning the essential difference between the edgewise moving helicopter rotor and the axial translating propeller was highlighted. In Chapters 4 and 7 the ability of the helicopter rotor to behave as a windmill during autorotation was introduced. As a finale, the windmill is examined in the light of previous chapters and the links between the windmill and the helicopter are highlighted.

## The basic types of windmill and their operation

Over the centuries the windmill has graced the landscape of many countries and if a closer look is made of their design then the skill and imagination of the millwright becomes apparent.[36-39]

The function of a windmill is to convert the energy of the wind into useful mechanical effort for turning heavy millstones or drawing water for drainage purposes. A windmill consists of a near horizontal axle, the *windshaft*, onto which several *sails* are attached via a hub. A large wheel, the *brake wheel*, is fitted to the windshaft and provides not only a mechanism to transfer the rotation of the windshaft to the internal axles in the mill, but also a means of stopping the mill by means of brake blocks pressing on its rim. The axis of the windshaft is turned to face directly into wind and the sails can then be driven by the wind, turning the hub and windshaft, and hence providing the mechanical effort.

It is of prime importance to keep the sails facing into wind, not only from an efficiency viewpoint, but also for safety of operation. The reason for this is that should the mill become "backwinded", where the wind comes from behind the mill, the axial load on the windshaft can cause it to withdraw from its bearings with the inevitable result of severe damage to the mill. The method of rotating the sails into wind distinguishes the various types of mill.

The *post mill* has the entire structure mounted on a vertical post. The upper structure, windshaft, hub and sails, are turned about the post by means of a long tailpole (see Fig. 15.1). Access to the mill interior is by a ladder which lies just above the tailpole and apart from providing access to the mill it adds a degree of weathercock stability to the mill aiding its rotation into wind.

The *tower mill* (Fig. 15.2) has a fixed cylindrical structure built of stone or brick. This is surmounted by a *cap* which carries the windshaft, hub and sails which can rotate about a vertical axis allowing the mill to be turned into wind. The

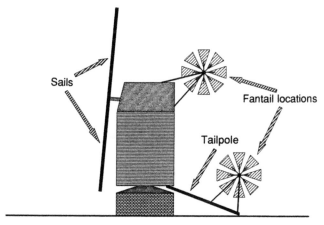

**Fig. 15.1.** Layout of a post mill.

cap rests on the *curb* which is a ring of wood or cast iron fitted to the top of the mill structure allowing it to turn about the vertical axis. Immediately below the cap rim is a rack attached to the curb and meshing with a pinion fitted to the cap. Rotation of the pinion causes the cap to rotate about a vertical axis as required and this is accomplished by means of the *luffing gear* which comprises an endless chain hanging from a wheel connected to the pinion and hanging down to ground level.

The *smock mill* is basically the same as a tower mill except that the main structure is built of wood and can be of a polygonal cross-section.

## Hub and sail attachment

The early mills used either two or four sails attached to the hub as opposite pairs. They were joined by large timbers of square cross-section known as *sail stocks*. These were then mortised to the end of the windshaft which is of square cross-section at this end, by means of an attachment known as a *poll end*. The sails, which are a framework of wood, have a main radial spar known as the *sail whip*, which is used to bolt the sail to the stock.

Later mills had a poll end made of metal but a further development by John Smeaton is a hub arrangement known as a *cross* which enabled any number of sails to be fitted. Each sail has an individual socket into which it is secured, not unlike the mountings for propeller blades. With this technique the sail whips had to be more substantial and became known as *sail backs*.

The ability of the cross principle for the hub in allowing flexibility in the choice of the number of sails confers an advantage over the poll end fixture, which can only support two or four blades. Should a sail become damaged, a poll end will be reduced to zero or two sails in order to keep a balanced windmill. (Two sails can produce approximately 60% of the power generated by four equivalent sails.) An odd number of sails does not allow a lesser number of sails to be used, hence the loss of a sail renders such a mill inoperative until the sail can be refitted. Use of six

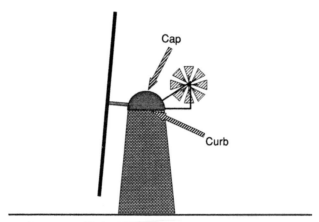

**Fig. 15.2.** Layout of a tower mill.

sails enables lower numbers of two, three, and four sails to be used giving great flexibility of operation. Some mills have used eight sails giving a high degree of sail area to extract wind energy. A second advantage of increasing the number of sails is in the smoothing effect on the power generation. During each revolution every sail will pass through the *tower shadow*, which is the region of relatively slow moving air caused by the mill structure blocking the wind flow. This will give a periodic pulsing to the power, and so with an increase in the number of sails, more sails will be unaffected by the tower shadow at any given moment and the mill will achieve a more even power extraction. While increasing the number of sails may at first sight be advantageous, the turbulence caused by the wakes off the sails has created problems. The difficulties of sail–vortex interaction were felt in windmills before helicopters encountered the same phenomenon.

## Sail construction and operation

The planform of a sail, shown in Fig. 15.3, consists of the whip which is the equivalent of a blade spar, morticed through which are the *sail bars* running in the chordwise direction and forming the equivalent of wing ribs. The trailing edge is formed by the *hemlath* which can be compared to the trailing edge tab of a rotor blade, and the leading edge has the *leading edge board* fitted. The sail surface which provides the lift, and hence the torque, comprises either cloth attached to the sail structure or rotating shutters mounted on chordwise pivots and attached to a *shutter bar* allowing them to open and close in unison, rather like a venetian blind.

Helicopter rotor thrust control is accomplished by means of variation of blade pitch angle, the blade area remaining constant. However, the windmill sail, because of its rigid attachment to the poll end or cross, cannot rotate about an axis aligned along the whip and so pitch cannot be used for thrust control. However,

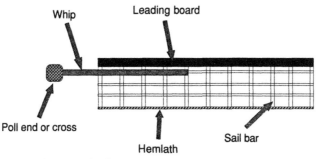

**Fig. 15.3.** Planform of a typical sail.

sail area can be adjusted and this is the means of controlling the mill as the wind velocity changes. In early mills the sail area was determined by the amount of cloth laid over the sail bars (*common sail*), and reefed according to various patterns as shown in Fig. 15.4. *Full sail* will extract the most energy from the wind while *first reef* has a very small amount of sail area placed close to the windshaft, and hence very little wind energy will be extracted. As the wind velocity changed the miller would have to trim the sails to suit, which required the mill to be stopped. Apart from being time consuming, this was a potential hazard since if a high wind was encountered unexpectedly, the sails would run out of control as the brakewheel was not strong enough to overcome the torque. Friction in the bearings and indeed in the millstones should they run dry (the grain acts as a lubricant) will generate much heat and in the past mills have been burnt to the ground by such occurrences. Using shutters, instead of cloth, allows the sail area to be readily adjusted. In 1789 Captain Stephen Hooper developed his *roller reefing sail*. The sail cloth was fitted to the sail frames in a manner rather like a set of roller blinds. The amount of cloth exposed to the wind was controlled by a *striking rod* which is fitted with a *spider* and slides along the axis of the windshaft and acts similarly to the actuator shaft connected to the spider on a helicopter tail rotor. In this way the sails can be adjusted while in motion.

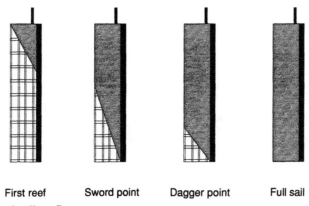

**Fig. 15.4.** Typical sail reefing patterns.

In 1772 Andrew Meikle, a Scottish millwright, invented the *spring sail* (Fig. 15.5). This attached the shutter bar to an adjustable spring placed near the hub and acting in a direction along the whip. Since the shutters rotate about an axle which does not lie along the centre of the shutter, as the wind speed increases the pressure on the shutters will begin to turn them out of the plane of the sail, and therefore the effective sail area will decrease. This will limit the torque from the sails and the principle can therefore act as a governor. The steady state conditions can be altered by adjusting the strength of the spring and whilst this allows a measure of control, the sails still require to be halted in order for the springs to be adjusted. In 1807 Sir William Cubitt patented his *self reefing* or *patent sail*. The principle is shown in Fig. 15.6 and the shutter bars in this system are controlled by a linkage consisting of the striking gear connected to a spider, as in Hooper's roller reefing system, and in a manner almost identical to the collective pitch control system of a tail rotor. The shutter position could then be altered by the miller while the sails were in motion. Although the torque derived from the wind is dependent on the sail area, the aerodynamic efficiency of the mill was improved by two features which are mirrored in a helicopter.

During the period 1600–1700 the radius of the sails increased and the advantages of lower disc loading used. Early sails had the whips placed at the half chord position and as the sail span and hence sail area increased, the aerodynamic loads grew accordingly, moving the sails towards potential aeroelastic instabilities such as flutter. With a whip at the half chord position, substantial torsional moments are obtained and because of this, the whip attachment, over the years, migrated towards the quarter chord position, used on helicopter rotor blades, and aligned the sail structure with the aerodynamic loads. The pitch axis of helicopter rotor blades is placed on or near to the blade quarter chord for similar reasons.

Early sail cross-sections were essentially flat (Fig. 15.7) giving a lift drag ratio of about three–see.[40] Later designs used a cambered construction with the leading board at an angle of about $22\frac{1}{2}°$ (nose down) to the chord line. This raised the lift drag ratio to about eight. Use of cambered aerofoils is now well established in helicopter blade design.

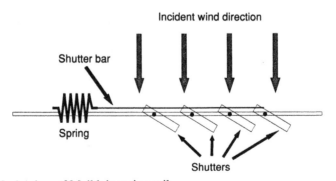

**Fig. 15.5.** Mechanism of Meikle's spring sail.

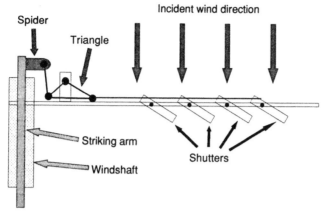

**Fig. 15.6.** Mechanism of Cubitt's patented self-reefing sail.

The efficiency of a helicopter rotor is improved, in hover particularly, from using washout or twist on the blades in order to spread the lift generation as uniformly as possible along the blade span and hence move it towards the ideal situation of the actuator disc. Windmill sails began to appear around 1600 with a twist incorporated, known as *weathering* (see Plate 15.2). As a rough guide, a windmill sail is set at a collective pitch of about 20° and is built with a weathering (washout) of about 15°.

## Yaw control—the fantail

The previous discussion is concerned with the windmill equivalent of the helicopter main rotor–what then of the tail rotor? One dominating theme in the opening remarks of this chapter is concerned with keeping the sails turned into wind and the inefficiency and safety considerations make this of prime importance. Early

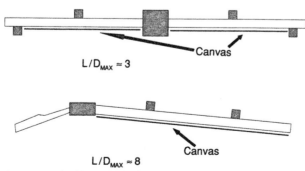

**Fig. 15.7.** Development of sail cross-sections.

mills used either the tail pole or luffing gear to turn the windshaft axis, however, this is a manual operation and is therefore entirely dependent on the vigilance of the miller. In 1745 Edmund Lee patented the *fantail*. This looks, to all intents and purposes, as a tail rotor fitted to the cap or tailpole (see Figs 15.1 and 15.2). It is an automatic device installed to sense changes in the wind direction and to turn the windshaft into wind. It effectively provides a weathercock stability similar to that provided by the tail rotor of a helicopter in the event of the fuselage being disturbed in yaw. It consists of a number of panels acting as rudimentary aerofoils and inclined equally to the vertical plane of rotation, as in the pitch setting of a tail rotor. As in the main sails, the pitch angle of the panels are fixed, but with the fantail the blade area is also fixed. The only variable is the rotational speed both in magnitude and direction.

As shown in Fig. 15.2, in a cap or smock mill the fantail is mounted on the cap directly opposite to the sails and its axle is connected with shafts and gears to the pinion and curb used to turn the sails into wind. A post mill can have the fantail positioned in two ways as shown in Fig. 15.1. The first location is on top of the roof and connected to a worm gear on the central post via a shaft running vertically down the leeward face of the mill itself. The second location is on a structure fitted to the tailpole and connected to a drive wheel, fixed to the tailpole and running along a circular track surrounding the entire mill.

As the wind direction changes, a component of wind velocity will be generated at the fantail in a direction parallel to its axle. This will induce a torque on the fantail sails causing it to rotate. Via the connecting gearing and shafts, this rotation is transferred to the mill structure causing it to turn towards the wind. As the mill becomes aligned to the new incident wind direction, the wind velocity component turning the fantail will diminish to zero, the fantail will stop rotating and the mill will halt in the correct position with the sails turned into the new wind direction.

The windmill has many features very similar to a helicopter, but used long before the first rotary wing airframe ever flew.

In the introduction, the essential differences between a helicopter rotor and a windmill were emphasised. However, in this final chapter it can be seen that similarities are also present, so the theories of the helicopter rotor are well founded in history with a distinguished pedigree.

**Plate 15.1.** The post mill at Brill, Buckinghamshire circa 1947.

**Plate 15.2.** Medmerry windmill at Selsey Bill, Sussex. Sail construction showing weathering.

# References

1. A. Gessow and G. C. Myers, *Aerodynamics of the Helicopter*. F. Ungar (1952).
2. M. C. Cheney, The ABC helicopter. *Journal of the American Helicopter Society*, October (1969).
3. D. L. Kohlman, *Introduction to V/STOL airplanes*. Iowa State University Press (1981).
4. I. C. Cheeseman, Circulation control and its application to stopped rotor aircraft. *Aeronautical Journal*, Vol. 72, July (1968).
5. R. M. Williams, Application of circulation control rotor technology to a stopped rotor aircraft design. *Vertica*, No. 1, pp. 3–15 (1976).
6. M. Rogers, *VTOL Military Research Aircraft*. Haynes Publishing (1989).
7. H. Glauert, *Airplane Propellers*, Vol. IV, *Aerodynamic Theory*. W. F. Durand. Springer (1935).
8. P. Brotherhood, An investigation in flight of the induced velocity distribution under a helicopter rotor when hovering. ARC RAE Report No. Aero 2212 (1947).
9. A. J. Landgrebe, The wake geometry of a hovering helicopter rotor and its influence on rotor performance. *American Helicopter Society Annual Forum* (1972).
10. W. E. Bennett and I. C. Cheeseman, The effect of the ground on a helicopter rotor in forward flight. ARC R&M 3021 (1957).
11. H. Glauert, *The Elements of Aerofoil and Airscrew Theory*. Cambridge University Press, 1926 (Reissued 1983).
12. W. Castles and J. H. De Leeuw, The normal component of the induced velocity in the vicinity of a lifting rotor and some examples of its application. NACA Report 1184, August (1952).
13. W. Z. Stepniewski and C. N. Keys, *Rotary Wing Aerodynamics*, Vol. I. Dover (1984).
14. R. Prouty, *Helicopter Aerodynamics*. PJS Publications (1985).
15. B. W. McCormick, *Aerodynamics of V/STOL Flight*. Academic Press (1967).
16. O. C. Zienkiewicz and R. L. Taylor, *The Finite Element Method*. McGraw-Hill (1989).
17. D. E. H. Balmford, R. E. Hansford and S. P. King, *Helicopter Dynamics*. Westland Helicopters Ltd Publication, March (1985).
18. R. W. White, A fixed frequency rotor head vibration absorber based upon GFRP springs. *Fifth European Rotorcraft Forum*, Amsterdam (1979).
19. T. M. Gaffney and R. W. Balke, Isolation of rotor induced vibration with the

Bell focal pylon nodal beam system. Paper 760892. Society of Automotive Engineers, *National Aerospace Engineering and Manufacturing Meeting*, San Diego, November (1976).

20. R. Jones, Control of helicopter vibration using the dynamic antiresonant vibration isolator. Society of Automotive Engineers, *National Aerospace Engineering and Manufacturing Meeting*, Los Angeles, October (1973).

21. S. P. King and A. Staple, Minimisation of helicopter vibration through active control of structural response. *AGARD Conference Proceedings 423*, October (1987).

22. G. M. Byham, An overview of conventional tail rotors. *Royal Aeronautical Society Conference on Helicopter Yaw Control Concepts*, February/March (1990).

23. R. R. Lynn, F. D. Robinson, N. N. Batra and J. M. Duhon, Tail rotor design. Part I—Aerodynamics. *Journal of the American Helicopter Society*, October (1970).

24. C. V. Cook, Tail rotor design and performance. *Vertica*, Vol. 2, pp. 163–181 (1978).

25. A. Brocklehurst, A significant improvement to the low speed yaw control of the Sea King using a tail boom strake. *Eleventh European Rotorcraft Forum*, London (1985).

26. R. J. Huston and C. E. K. Morris, A note on a phenomenon affecting helicopter directional control in rearward flight. *Journal of the American Helicopter Society*, October (1970).

27. J. W. Leverton, J. S. Pollard and C. R. Wills, Main rotor wake/tail rotor interaction. *Vertica*, Vol. 1, pp. 213–221 (1977).

28. R. Mouille, The fenestron shrouded tail rotor of the SA 341 Gazelle. *Journal of the American Helicopter Society*, October (1970).

29. A. L. Winn and A. H. Logan, The MDHC NOTAR™ system. *Royal Aeronautical Society Conference on Helicopter Yaw Control Concepts*, February/March (1990).

30. R. Prouty, *More Helicopter Aerodynamics*. PJS Publications (1988).

31. W. Gunston, *AH-64 Apache (Combat Aircraft Series)*. Osprey (1986).

32. K. W. Burton and D. V. Ellis, Aeroelastic instabilities of rotor blades and tail rotor instabilities. Westland Helicopters Ltd, Yeovil, March (1977).

33. L. A. Pars, *A Treatise of Analytical Dynamics*. Heinemann (1965).

34. R. P. Coleman and A. M. Feingold, Theory of self-excited mechanical oscillations of helicopter rotors with hinged blades. NACA Report 1351 (1958).

35. W. G. Bousman and D. L. Winkler, Application of the moving block analysis. *AIAA Dynamics Conference*, Atlanta, GA, April (1981).

36. R. Wailes, *The English Windmill*. Routledge Kegan & Paul (1954).

37. S. Beedell, *Windmills*, David & Charles (1975).

38. F. Stokhuyzen, *The Dutch Windmill*. Universe Books (1954).

39. J. Vince, *Discovering Windmills*. Shire Publications (1984).

40. J. M. Drees, Blade twist, droop snoot and forward spars. *American Helicopter Society—Vertiflite*, Vol. 22, No. 5 (1976).

# Bibliography

To enable the reader to proceed into any topic further, the following list of textbooks, reports and papers is provided. A vast amount of reading matter is available and this list can only scratch the surface. However, it is intended that this will provide a useful starting point.

As can be seen from the references and this list, certain sources of information are of great importance. Conferences, meetings and symposia organised by The Royal Aeronautical Society and The American Helicopter Society and their associated journals must be highlighted. Other sources are the proceedings of the annual European Rotorcraft Forum, and the relevant journals of the American Institute of Aeronautics and Astronautics.

The index provided by the International Aerospace Abstracts provides a valuable general aviation database of technical publications.

## General interest and historical

M. J. Taylor, *History of the Helicopter*, Hamlyn (1984).

J. Schafer, *Basic Helicopter Maintenance*. International Aviation Publishers (1980).

W. J. Boyne and D. S. Lopez (Editors), *Vertical Flight: The Age of the Helicopter*. Smithsonian Institution Press (1984).

G. Townson, *Autogiro: The Windmill Plane*. Aero Publishers (1985).

D. N. James, *Westland Aircraft Since 1915*. Putnam (1991).

Lt Cdr J. M. Milne, *Flashing Blades Over the Sea: The Development and History of Helicopters in the Royal Navy*. Maritime Books, Liskeard.

P. Beaver, *Modern Military Helicopters*. Patrick Stephens (1987).

Capt. E. Brown, *The Helicopter in Civil Operations*. Granada (1981).

W. Gunston, *Helicopters of the World*. Temple Press (1983).

S. Dalton, *The Miracle of Flight*. Sampson Low (1977).

H. Penrose, *Adventure with Fate*. Airlife (1984).

R. Gardner and R. Longstaff, *British Service Helicopters: A Pictorial History*. Robert Hale (1985).

J. Fay, *The Helicopter, History, Piloting and How it Flies*. David & Charles (1987).

D. Wragg, *Helicopters at War: A Pictorial History*. Robert Hale (1983).

J. Hobbs, *Bristol Helicopters: A Tribute to Raoul Hafner*. Frenchay Publications, Bristol (1984).

W. Gunston and M. Spick, *Modern Fighting Helicopters*. Salamander (1986).

M. J. Ingham, *To the Sunset Bound: 1987—50th Anniversary of Scheduled Air*

*Services to the Isles of Scilly.* Air-Britain (Historians) Ltd, Tonbridge (1987).

F. W. Free, Russian helicopters. *Aeronautical Journal*, September, pp. 767–785 (1970).

*A History of British Rotorcraft 1866–1965.* Westland Helicopters Ltd, Yeovil (1968).

A. Stepan, The pressure jet helicopter. *Journal of The Royal Aeronautical Society*, Vol. 62, February (1958).

M. Spick, *MIL MI-24 Hind.* Osprey (1988).

M. Lambert (Editor), *Janes All the World's Aircraft.* Published annually.

J. Everett-Heath, *British Military Helicopters.* Arms & Armour Press (1986).

M. Lloyd, *The Guinness Book of Helicopter Facts and Feats.* Guinness Books (1993).

## Aerodynamics

R. Hafner, The domain of the helicopter. *Journal of The Royal Aeronautical Society*, Vol. 58, October (1954).

J. P. Jones, The helicopter rotor. *Aeronautical Journal*, November (1970).

F. D. Harris, Rotary wing aerodynamics—historical perspective and important issues. *National Specialists' Meeting on Aerodynamics and Aeroacoustics*, American Helicopter Society, February (1987).

A. R. S. Bramwell, A note on the static pressure in the wake of a hovering rotor. Research Memorandum, City University, London, Aero 73/3 (1973).

I. A. Simons, R. E. Pacifico and J. P. Jones, The movement, structure and breakdown of trailing vortices for a rotor blade. ARC 28993, April (1967).

C. V. Cook, The structure of the rotor blade tip vortex. *AGARD Conference Proceedings* CP-111 (1972).

M. P. Scully, A method of computing helicopter vortex wake distortion. M.I.T. ARSL TR 138-1, June (1967).

J. D. Kocurek and J. L. Tangler, A prescribed wake lifting surface hover performance analysis. *32nd American Helicopter Society Annual Forum,* May (1976).

D. S. Jenney, J. R. Olson and A. J. Landgrebe, A reassessment of rotor hovering performance prediction methods. *Journal of the American Helicopter Society*, No. 1 (1968).

C. V. Cook, Rotor performance prediction in hover. Westland Helicopters Ltd, Research Paper 357, November (1968).

C. V. Cook, Induced flow through a helicopter rotor in forward flight. Westland Helicopters Ltd, Research Paper 374, January (1970).

P. Brotherhood, Flow through a helicopter rotor in vertical descent. ARC Report No. 2735 (1949).

P. Brotherhood and W. Stewart, An experimental investigation of the flow through a helicopter rotor in forward flight. ARC Report No. 2734 (1949).

K. W. Mangler and H. B. Squire, The induced velocity field of a rotor. ARC R&M 2642, May (1950).

H. H. Heyson and S. Katzoff, Induced velocities near a lifting rotor with non-uniform disc loading. NACA Report 1319 (1958).

M. L. Mil, Helicopter calculation and design, Vol. I, aerodynamics. NASA TT F-494 (1967).

Lt Cdr A. S. Ellin, An in-flight experimental investigation of helicopter main rotor/tail rotor interactions. Ph.D. thesis, University of Glasgow, April (1993).

I. C. Cheeseman and M. M. Soliman, A high speed edgewise rotor using circulation control only in the reverse flow area. *17th European Rotorcraft Forum* (1991).

T. S. Beddoes, Two and three dimensional indicial methods for rotor dynamic airloads. *American Helicopter Society/National Specialist Meeting on Rotorcraft Dynamics*, Arlington, November (1989).

T. S. Beddoes, Three dimensional separation model for arbitrary planforms. *47th American Helicopter Society Annual Forum*, Phoenix, May (1991).

H. Ashley, R. L. Halfman and R. L. Bisplinghoff, *Aeroelasticity*. Addison-Wesley (1955).

E. H. Dowell *et al.*, *A Modern Course in Aeroelastics*. Sijthoff & Noordhoff (1980).

## Fuselage aerodynamics

J. Seddon, *Basic Helicopter Aerodynamics*. Blackwell (1990).

S. F. Hoerner, *Fluid Dynamic Drag*. Published by the author (1965).

## Acoustics

J. W. Leverton, The sound of rotorcraft. Westland Helicopters Ltd, Research Paper RP 390, December (1970).

A. C. Pike, Validation of high frequency airload calculations using full scale flight test acoustic data. *American Helicopter Society/Royal Aeronautical Society Specialists' Meeting on Rotorcraft Acoustics and Rotor Fluid Dynamics*, Philadelphia, October (1991).

## Dynamics and vibration

G. Reichart, Basic dynamics of rotors, control and stability of rotary wing aircraft, aerodynamics and dynamics of advanced rotary wing configurations.

AGARD Lecture Series 63 on Helicopter Aerodynamics and Dynamics (1973).

J. C. Houbolt and G. W. Brooks, Differential equations of motion for combined flapwise bending, chordwise bending and torsion of twisted non-uniform rotor blades. NACA Report 1346 (1958).

R. E. D. Bishop and D. C. Johnson, *The Mechanics of Vibration*. Cambridge University Press (1960).

G. Isakson and J. G. Eisley, Natural frequencies in coupled bending and torsion of twisted, rotating and non-rotating blades. NASA CR-65 (1964).

R. W. Balke *et al.*, Tail rotor design. Part II, structural dynamics. *Journal of the American Helicopter Society*, October (1970).

R. E. Hansford and I. A. Simons, Torsion flap lag coupling on helicopter rotor blades. *Journal of the American Helicopter Society*, No. 3 (1973).

M. L. Mil, Helicopters—calculations and design, Vol. II, vibrations and dynamic stability. NASA TT F-519 (1968).

G. T. S. Done, A simplified approach to helicopter ground resonance. *Aeronautical Journal*, May (1974).

A. R. S. Bramwell, An introduction to air resonance. ARC Report Memorandum No. 3777 (1975).

D. E. H. Balmford, Vibration testing of helicopters. *Aeronautical Journal*, Vol. 74, August (1970).

J. P. Jones, Helicopter vibrations. *Journal of The Royal Aeronautical Society*, Vol. 64, December (1960).

J. B. Hunt, *Dynamic Vibration Absorbers*. Mechanical Engineering Publications Ltd, London (1979).

## General design calculations

W. Z. Stepniewski and C. N. Keys, *Rotary Wing Aerodynamics*, Vol. II. Dover (1984).

M. N. Tishchenko, A. V. Nekrasov and A. S. Radin, *Helicopters—Selection of Design Parameters*. Mashinostroyeniye Press, Moscow (1976). [Translated by W. Z. Stepniewski and W. L. Metz, Contract No. AVRADCOM NAS2-10062, April (1979).]

## Concept and evolution of a particular helicopter

The following two papers describe the original concept and design of the Westland WG13 helicopter. In addition to the technical aspects of a helicopter's evolution, a vitally important factor is the requirements of the customer. These two papers tell the story of the WG13 project during its formative years and describe the ramifications of changing requirements and in particular how these propagate throughout the entire aircraft.

R. G. Austin, The concept of the WG13. *Aeronautical Journal*, January (1974).

V. A. B. Rogers, The design of the WG13. *Aeronautical Journal*, January (1974).

## Mathematics and computing

W. H. Press *et al.*, *Numerical Recipes*. Cambridge University Press (1986).

M. R. Spiegel, *Laplace Transforms*. Schaum Publishing (1965).

M. R. Spiegel, *Vector Analysis*. Schaum Publishing (1959).

D. A. Wells, *Lagrangian Dynamics*. Schaum Publishing (1967).

H. Dwight, *Tables of Integrals and Other Mathematical Data*. Macmillan (1961).

## *Detailed helicopter topics*

A. R. S. Bramwell, *Helicopter Dynamics*. Edward Arnold (1976).

W. Johnson, *Helicopter Theory*. Princeton University Press (1980).

R. Prouty, *Helicopter Performance, Stability, and Control*. Robert E. Krieger Publishing (1990).

Boeing Vertol Chinook tandem rotor helicopter. Note the rear rotor pylon and the rear loading facility with a ramp door. As can be seen, the rear loading bay with a ramp door is ideal for transport operations, but the aerodynamic drag of the fuselage suffers because of the bluff rear end.

# Index

Printed and bound by CPI Group (UK) Ltd, Croydon, CR0 4YY

03/10/2024

01040331-0010